Alzheimer's Dementia

CONTEMPORARY ISSUES IN BIOMEDICINE, ETHICS, AND SOCIETY

Alzheimer's Dementia: Dilemmas in Clinical Research, edited by **Vijaya L. Melnick** and **Nancy N. Dubler,** *1985*

Feeling Good and Doing Better, edited by **Thomas H. Murray, Willard Gaylin,** and **Ruth Macklin,** *1984*

Ethics and Animals, edited by **Harlan B. Miller** and **William H. Williams,** *1983*

Profits and Professions, edited by **Wade L. Robison, Michael S. Pritchard,** and **Joseph Ellin,** *1983*

Visions of Women, edited by **Linda A. Bell,** *1983*

Medical Genetics Casebook, by **Colleen Clements,** *1982*

Who Decides?, edited by **Nora K. Bell,** *1982*

The Custom-Made Child?, edited by **Helen B. Holmes, Betty B. Hoskins,** and **Michael Gross,** *1981*

Birth Control and Controlling Birth, edited by **Helen B. Holmes, Betty B. Hoskins,** and **Michael Gross,** *1980*

Medical Responsibility, edited by **Wade L. Robison** and **Michael S. Pritchard,** *1979*

Contemporary Issues in Biomedical Ethics, edited by **John W. Davis, Barry Hoffmaster,** and **Sarah Shorten,** *1979*

UAHJN
TELEPEN

UAHJN

Alzheimer's Dementia

Dilemmas in Clinical Research

Foreword by

Robert N. Butler

Edited by

Vijaya L. Melnick and Nancy N. Dubler

Humana Press · Clifton, New Jersey

Crescent Manor
PO Box 2148
Clifton, NJ 07015

Printed in the United States of America

Library of Congress Cataloging in Publication Data
Main entry under title:

Alzheimer's dementia.

(Contemporary issues in biomedicine, ethics, and society)
Expanded and updated papers from a conference held Nov. 1981, sponsored by the National Institute on Aging.
Includes bibliographies and index.
1. Alzheimer's disease—Congresses. I. Melnick, Vijaya L. II. Dubler, Nancy N. III. National Institute on Aging. IV. Series. [DNLM: 1. Alzheimer's Disease—congresses. WM 220 A477]
RC523.A385 1985 618.97′683 85–11740
ISBN 0–89603–067–9

Permission to reprint "Clinical Research in Senile Dementia of the Alzheimer's Type: Suggested Guidelines Addressing the Ethical and Legal Issues" from the Journal of the American Geriatrics Society is gratefully acknowledged.

The views expressed in this paper do not necessarily represent the views of the National Institute on Aging or its parent agencies, the National Institutes of Health and the Department of Health and Human Services. Official support or endorsement of this paper by these agencies is not intended nor should it be inferred.

Contents

ETHICAL ISSUES IN THE CARE OF THE PATIENT INVOLVED IN ALZHEIMER'S DISEASE RESEARCH
Edward W. Campion

Part 3 HISTORICAL, LEGAL, AND ETHICAL BACKGROUND

RESEARCH OBJECTIVES AND THE SOCIAL STRUCTURING OF THE RESEARCH ENTERPRISE: AN HISTORICAL AND ETHICAL PERSPECTIVE
Harry Yeide, Jr.

RESEARCH ON SENILE DEMENTIA OF THE ALZHEIMER'S TYPE: ETHICAL ISSUES INVOLVING INFORMED CONSENT
Christine Cassel

LEGAL ISSUES IN RESEARCH ON INSTITUTIONALIZED DEMENTED PATIENTS
NANCY NEVELOFF DUBLER

ISSUES OF EQUITY IN THE SELECTION OF SUBJECTS FOR EXPERIMENTAL RESEARCH ON SENILE DEMENTIA OF THE ALZHEIMER'S TYPE
HARRY R. MOODY

PART 5 COMPETENCY TO GIVE CONSENT

COMPETENCY TO CONSENT TO RESEARCH
BARBARA STANLEY

ASSURING ADEQUATE CONSENT: SPECIAL CONSIDERATIONS IN PATIENTS OF UNCERTAIN COMPETENCE
Alan Meisel

ASSESSMENT OF COMPETENCE TO GIVE INFORMED CONSENT
Allen R. Dyer

Part 6 PROXY AND DERIVED CONSENT

AUTONOMY AND PROXY CONSENT
Bruce L. Miller

Foreword

The National Institute on Aging (NIA) has historically been concerned with the protection of human subjects. In July 1977, the NIA sponsored a meeting to update and supplement guidelines for protecting those participating in Federal research projects. Although the basic guidelines had been in effect since 1966, it had been neglected to include the elderly as a vulnerable population. In November 1981, the NIA organized a conference on the ethical and legal issues related to informed consent in senile dementia cases.

The present volume offers the latest and best thinking on Alzheimer's Dementia to have emerged from the dialog that was first embarked upon at the NIA meeting. Indeed, the issues and concerns it treats now seem even more relevant than they appeared historically because of the vastly greater awareness in the community of the entire spectrum of problems Alzheimer's disease confronts us all with.

Our interest and concern is both humanitarian and self-serving. Clearly older people must be protected from inappropriate research and careful attention must be paid to the circumstances under which research is conducted on those older persons who have given anything less than full consent. It is equally necessary, however, for the research enterprise to be protected so that today's elderly and those of the future can benefit from the fruits of research.

The focus of the discussion that follows—the dementing disorders of later life—complicates the issue even further since the diseases themselves affect the capacity of the individual to give informed consent. How are we to design sound protocols—from both a scientific and an ethical perspective—for research related to understanding or treating the dementing illnesses, but also for research not related to dementia that nonetheless might be conducted in patients suffering from such diseases? How clear is our knowledge base concerning the degree to which the capacity for informed consent is affected? What are the possibilities for consent being given long prior to development of the disease (predisease consent) or by a family member or advocate (proxy consent)?

Perhaps it is necessary to remind ourselves of the profound nature and devastating impact of the dementias. Senile dementia is a disease or a set of diseases that destroy the individual and, more often than not, the emotional, social, and financial lives of that individual's family. In terms of their impact on the culture as a whole, these conditions often produce abhorrence and negativism toward the elderly by those who are fearful of senility. The costs to individuals and society-at-large are staggering. There are now more than 1.3 million persons in nursing homes, including 1 million over age 65. Perhaps as many as half of the elderly in nursing homes suffer from some degree of senile dementia, although they may have other disease in addition.

At present, the population rises at a rate of 1600 older persons per day, or some 600,000 per year. It is not surprising, then, in this century of old age, that Lewis Thomas has called senile dementia the "disease of the century." In the not-too-distant future, the post-World War II "baby boom" generation will grow gray, adding to the impact of declining mortality rates on the US demographic profile. Obviously, new forms of care, but most of all new knowledge, will be essential to offset the expanding numbers of people suffering from senile dementia.

America is not alone in experiencing this demographic revolution. There is a worldwide graying of nations and a concomitant anticipation of the potential impact of senile dementia in the near future. The World Health Organization's Advisory Committee on Medical Research has identified senile dementia as an issue of global concern.

Sound scientific research on the dementias of aging requires an interdisciplinary approach. So, too, does the development of ethical guidelines. We must have clinicians, investigators, lawyers, ethicists, and older persons themselves join forces in developing appropriate research protocols. It is essential that this multidisciplinary team also include family members of persons suffering from senile dementia. I am pleased that the Alzheimer's Disease and Related Disorders Association—a voluntary organization of victims, families, and friends and a new development on the national scene—has chosen to play an active role in this discussion.

The National Institute on Aging takes some pride in the strides we have made in encouraging new research on senile dementia. Together with the National Institute of Neurological and Communicative Disorders and Stroke, the National Institute of Mental Health, and the National Institute of Allergy and Infectious Diseases, we have sponsored a major research effort to this end. We have initiated epidemiological studies to provide us with a greater knowledge of the incidence and prevalence of the dementias of aging.

We have also developed a new grant mechanism, the Teaching Nursing Home, that will begin to open the door to a previously sequestered population. Many of our public policies have led to the exclusion of victims of senile dementia of the Alzheimer's type from state mental hospitals and academic hospitals. Many are in nursing homes. It is important that academic investigators enter the nursing home, not only to improve health care, but to learn more about senile dementia and the various other conditions found among the elderly.

In our efforts to encourage research in nursing homes, we have been cognizant of the dangers of inappropriately using this vulnerable population for scientific gain. We have also been among the first to focus attention on the need to protect elderly research subjects. This fine volume, edited by Dr. Vijaya Melnick and Ms. Nancy Dubler, should help all of us, professionals and lay persons, in this country and abroad, to develop the kind of guidelines that will ultimately lead us to the scientific answers that we seek—within an ethical framework.

Robert N. Butler, MD

Preface

Senile Dementia of the Alzheimer's Type (SDAT) is a devastating condition affecting millions of American patients and their families. It is a disease that inexorably diminishes and ultimately destroys the ability of individuals to comprehend their condition and survive in their prior environment. It requires Herculean physical and emotional efforts from family members who choose to keep patients at home. It may require the impoverishment of families who select institutionalization for the patient. Grief, suffering, guilt, and endless mourning are the constant companions of spouses, siblings, and children of these patients.

The costs to society are equally weighty. Long-term institutional care now approaches forty thousand dollars per year. The costs of lost productivity of patients and families are incalculable.

At present, SDAT is difficult to diagnose and impossible to cure. Prevention is a future dream. However, certain avenues of research are promising. Society must grapple with how this research can be pursued, given the particular disabilities of these potential research subjects.

In November 1981, the National Institute on Aging (NIA) sponsored a conference on "Senile Dementia of the Alzheimer's Type (SDAT) and Related Diseases: Ethical and Legal Issues Related to Informed Consent." The conference was convened to explore the values, conflicts, and competing interests that must be accommodated if research is to continue on the pathophysiological processes and psychosocial aspects of dementia.

The papers presented at the conference and the discussions that followed pointed to a clear need for research to go forth. However, what type of research may be acceptable, in the context of present Federal regulations governing research on human subjects, was not at all clear. Questions were raised as to:

1. Who might or should be permitted to speak on behalf of or as a substitute for the patient, if the patient is not capable of giving competent informed consent? SDAT patients most often are in a progressive state of declining competence.
2. How could protocols be designed, in the most ethical context, that are scientifically acceptable and that will per-

mit research to further the knowledge on the understanding and treatment of dementing illnesses?

3. Can patients afflicted with such diseases ever grant effective informed consent to participate in a research protocol?
4. Can patients ever be morally admitted to protocols in the absence of their own consent or the adequate consent of others? How is the role of the family and close friends defined in this context?
5. What are the possibilities for obtaining consent prior to the development of the illness, or to record consent at an early stage of the illness for later research intervention?
6. What are the additional constraints that are placed on researchers attempting to address the complex problems of dementing illnesses?
7. Can society make a claim, however minimal, on the supposed altruism of patients if the intervention poses minimal risk and the possible benefit is great?

These were the major questions posed to participants at the conference. The articles especially expanded and updated for this volume were created in response to those questions.

Whereas the tone and tenor of the discussion was clearly in support of continued research, all agreed that the legal and moral uncertainties surrounding continued efforts are substantial. Research on SDAT patients must confront society with uncomfortable choices. One choice is to disregard previously clear prerequisites for participation in research by ignoring the requirements for individual contemporaneous informed consent. Another option is to define and develop new moral and legal principles to support third party consent; these alternatives could be mandated in regulation. A third path is to proceed as we now are with gerrymandered logic and procedures, which may compromise patient rights and often place investigators in morally compromising situations. It is the hope of the editors that this volume and the guidelines developed as a result of the analyses it presents will aid in devising new routes through these thorny thickets.

The guidelines, with which the volume ends, are not regulations. They do not create law. They attempt, given the present state of our legal and moral analysis, to direct researchers to more clear routes of planning and action.

We hope that these articles and guidelines will be of both theoretical interest and practical importance.

Vijaya L. Melnick
Nancy N. Dubler

CONTRIBUTORS

ROBERT N. BUTLER • *Department of Geriatrics and Adult Development, Mt. Sinai Medical Center, New York, New York*

EDWARD W. CAMPION • *Geriatrics Unit, Massachusetts General Hospital, and Spaulding Rehabilitation Hospital, Boston, Massachusetts*

CHRISTINE CASSEL • *Department of Geriatrics and Adult Development, Mt. Sinai Medical Center, New York, New York*

NANCY NEVELOFF DUBLER • *Department of Social Medicine, Montefiore Medical Center, Bronx, New York*

ALLEN R. DYER • *Department of Psychiatry, Duke University Medical Center, Durham, North Carolina*

STEVEN H. FERRIS • *Department of Psychiatry, New York University Medical Center, New York, New York*

BETTY GORDON • *Department of Psychiatry, New York University Medical Center, New York, New York*

PAMELA B. HOFFMAN • *Department of Geriatric Medicine, New Hyde Park, New York*

ANDREW JAMETON • *Department of Medical Jurisprudence and Humanities, University of Nebraska Medical Center, Omaha, Nebraska*

ROBERT KATZMAN • *Albert Einstein College of Medicine, Bronx, New York*

ROBERT J. LEVINE • *Department of Medicine, Yale University School of Medicine, New Haven, Connecticut*

LESLIE S. LIBOW • *Department of Geriatrics and Adult Development, The Jewish Home and Hospital for Aged, New York, New York*

CHARLES R. MCCARTHY • *Office for Protection from Research Risks, National Institutes of Health, Bethesda, Maryland*

MARTIN MCCARTHY • *Department of Psychiatry, New York University Medical Center, New York, New York*

ALAN MEISEL • *Department of Law, University of Pittsburgh School of Law, Pittsburgh, Pennsylvania*

VIJAYA L. MELNICK • *Department of Biology and the Center of Applied Research and Urban Policy, University of the District of Columbia, Washington, DC*

BRUCE L. MILLER • *Medical Humanities Program, Michigan State University, East Lansing, Michigan*

HARRY R. MOODY • *Brookdale Center on Aging, Hunter College, New York, New York*

NANCY C. PASCHALL • *Patient's Rights and Advocacy Section, Mental Health Services Development Program, NIMH, Rockville, Maryland*

HILDA PRIDGEON • *Alzheimer's and Related Diseases Association, Bloomington, Minnesota*

RICHARD M. RATZAN • *University of Connecticut School of Medicine, Farmington, Connecticut*

BARRY REISBERG • *Department of Psychiatry, University Medical Center, New York, New York*

BARBARA STANLEY • *Department of Psychiatry, Wayne State University School of Medicine, Detroit, Michigan*

LANCE TIBBLES • *Capital University Law School, Columbus, Ohio*

ALAN WEISBARD • *Cardoza Law School, Yeshiva University, New York, New York*

HARRY YEIDE, JR. • *Department of Religion, George Washington University, Washington, DC*

Part 1

Legal and Science Background

Current Frontiers in Research on Alzheimer's Disease

Robert Katzman

Risks and Benefits of Research

Implicit in the discussion of bioethical issues in regards to Alzheimer's disease is the question of the relationship of possible risk and expected benefit from further investigations. The most significant benefit to the person, the family, and to society would be a breakthrough in our understanding of a disease process that is at present not treatable and not preventable—a disease that is as malignant in its own way as cancer.

We do not understand the causes or the etiology of Alzheimer's disease, but within the past 20 years, remarkable progress has been made in defining the disease clinically and in understanding what happens in the brain. Alzheimer's disease, as a characteristic clinical–pathological entity affecting a small number of individuals in the presenium, was readily accepted following the description of the pathological changes in cerebral cortex and hippocampus by Alois Alzheimer in 1907. The consensus that the same disorder is in fact the major cause of senile dementia, and therefore one of the most important and frequent diseases today from a public health point of view, has only developed in the past 15 years.

The key to this recognition of the role of Alzheimer's disease as a major affliction of the seniors has been clarification of the nature of senile dementia, the recognition that senile dementia is not part of "normal aging," that it is a specific symptom complex—clinically separable from affective disorders—a symptom complex always associated with diseases that affect the brain either primarily or secondarily. Senile dementia has been reported to be associated with over 50 diseases, but just over 50% of cases are caused by Alzheimer's disease alone, 20% by vascular diseases resulting in multiple stroke (so called multi-infarct dementia), and 10 to 15% by a mixture of Alzheimer's disease and vascular dementia. The remaining 15 to 20% of cases are associated with 45 or more other disorders. Though some of these are at present progressive, irreversible brain diseases (e.g. Huntington's disease, Pick's disease, progressive supranuclear palsy), others are quite treatable—including vitamin deficiencies (e.g. Korsakoff's disease with thiamine deficiency, pellagra with niacin deficiency, vitamin B12 deficiency), systemic disorders such as endocrinopathies (e.g. hypothyroidism), infections (e.g. neurosyphilis), hydrocephalus, and brain tumors. Thus, a thorough workup of every patient with dementia is essential in order to identify treatable disorders, but also in order to provide a rational basis for management of those conditions for which we do not have a specific treatment.

Testing and Measurement of Alzheimer's Dementia

Having recognized that not all elderly individuals with dementia need have Alzheimer's disease, one may inquire further as to relationship of the microscopic changes seen in the brain post mortem and the degree of dementia manifested during life. This question was addressed in the late 1960s by Blessed, Tomlinson, and their colleagues in a classic prospective study carried out in Newcastle. In this study, Blessed evaluated older patients to determine whether dementia was present, and if present, its degree as manifested by a score on a mental status test and by a score on a functional disability scale. Seventy eight of these individuals died and were autopsied. Tomlinson counted the number of neuritic plaques (disordered tangles of neurons) in microscopic sections of specified areas of cerebral cortex and demonstrated that this number correlated well with the degree of dementia measured both by the functional scale ($r = 0.7$) and the mental status test score ($r = 0.6$). These investigations also demonstrated that the same degree of dementia occurred in patients with vascular disease of the brain when more than 50 to 100 grams of brain tissue were destroyed by strokes.

The correlation of clinical and postmortem findings in Alzheimer's disease has recently been further extended by demonstration of a similar relationship between loss of the enzyme choline acetyl transferase and dementia scores obtained during life. This correlation of clinical findings and pathological and biochemical changes suggests that detailed investigations of such changes should be fruitful. Indeed, major advances in our understanding of Alzheimer's disease have resulted from studying what is happening in the brain and relating this to clinical findings.

In 1964, Terry in the United States and Kidd in England described the electron microscopic features of the neurofibrillary tangle and the neuritic plaque. Their pictures were indeed dramatic. The neurofibrillary tangle consists of arrays of thousands of submicroscopic fibrils, which are pairs of abnormal filaments wound around each other in a spiral or helical pattern, the bihelical filament. Each individual filament contributing to the bihelical pair is only 100 Ångstroms in width. These filaments differ from normal filaments present in neurons, filaments that are also 100 Angstroms in width, but which are linear, not twisted, and which have numerous side arms not present in the abnormal bihelical filament.

The neuritic plaque consists of degenerating nerve endings, but surprisingly in many plaques, these endings, although degenerating, have intact membranes, sometimes even with intact synaptic thickenings, complete with presynaptic vesicles that are believed to store neurotransmitters. There is also a proliferation of glial cells (the supporting cells in the brain) within this neuritic plaque. The central core of the plaque contains a mesh work of another fibrous protein that has the appearance of amyloid protein, a form of protein often found in chronic diseases elsewhere in the body. In some instances, amyloid proteins represent the result of an immunological reaction to the disease state, whereas in other instances, they represent a particular pattern of protein breakdown in local tissue.

Fibrous Proteins Changes

Thus, changes in fibrous proteins are an important part of the Alzheimer's brain changes. What is the molecular nature of these proteins? This is difficult to study since the Alzheimer pathology, in particular the bihelical filament, is uniquely human and does not occur in animals. In order to study the protein chemistry of this fibrous material, one must apply biochemical and immunological techniques to brain tissue obtained at autopsy. Major advances have been made in the past several years because of the development of new monoclonal antibodies, specific antibodies that enable one to begin to identify the molecular nature of the protein and that may provide a basis for isolating the protein in a fashion suitable eventually for analysis. The bihelical filamentous protein has been difficult to isolate and characterize chemically, in part because of its insolubility. But this very insolubility may provide clues to its nature. Is the bihelical filament protein in normal protein secondarily altered and polymerized? Do the bihelical filaments accumulate because of overproduction or because they cannot be broken down?

Several laboratories have now obtained monoclonal antibodies to paired helical filament preparations. These antibodies react with neurofibrillary tangles in Alzheimer brains, but do not react with other normal nerve cells or other proteins of normal nerve cells, suggesting, but not proving, that this protein may be novel protein not expressed by the mature neuron. Such a finding, if true, might be an important clue to the etiology of Alzheimer's disease.

Monoclonal Antibodies

Similarly, the availability of various monoclonal antibodies should permit investigators to look at the molecular nature of the amyloid in the neuritic plaque. Does it represent a blood constituent and therefore indicate that an immune type of reaction has occurred? Does it represent a breakdown product of a normal brain protein? Congophilic angiopathy, reflecting the presence of amyloid sometimes occurs in blood vessels in the Alzheimer brain; is this fibrous protein analogous to neuritic plaque amyloid?

At present, work on the abnormal fibrous proteins of Alzheimer's disease can be continued using autopsy brain tissue. One can conceive, however, the situation in which major advances are made in determining the nature of the protein, following which specific questions about the presence or absence of, for example, an enzyme accounting for the accumulation of these abnormal filaments. This enzyme may be particularly labile and therefore could not be studied in post mortem tissue; brain biopsy material might be necessary. At this stage, would this procedure be justified?

Neurotransmitter Systems

Another major area of advance has been in the discovery of abnormalities in specific neurotransmitter systems in the brain. This is particularly exciting because it affords the possibility of treatment of symptoms during life and because it also points the way to identification of neuronal systems with important behavioral aspects whose existence had not been suspected in the past. Ten neurotransmitters have been studied so far in the Alzheimer brain. For eight of these transmitters, there were either no changes or small, but inconsistent, changes observed in postmortem Alzheimer's brains compared to brains of age-matched normals. Two transmitters, however, have shown quite consistent changes. Choline acetyltransferase (CAT), the enzyme required for the biosynthesis of the neurotransmitter acetylcholine, is decreased from 50 to 90% in Alzheimer brains. One neuropeptide, somotostatin, is also decreased by about 50% in the cortex of Alzheimer's brains. This change in choline acetyltransferase occurs in brains in which the muscarinic receptor in the cerebral cortex, the receptor upon which acetylcholine acts, is present in essentially normal amounts in Alzheimer brains. Moreover, the loss of choline acetyltransferase observed at autopsy in Alzheimer brains roughly parallels the degree of dementia measured during life. Thus, these changes in the cholinergic system, and probably also somotostatin, appear to be fairly specific to the Alzheimer brain.

A surprising finding has been that most of the choline acetyltransferase in the cerebral cortex is not found in nerve cell bodies, but rather in nerve endings in the cerebral cortex. These cholinergic terminals arise from nerve processes projecting from a subcortical nucleus to the cerebral cortex. The cell bodies in this cholinergic projection system lie ventral to the basal ganglia in the region of the substantia innominata or nucleus basalis of Meynert. It should be noted that the discovery of the presence of the cholinergic projection system from nucleus basilis to cerebral cortex resulted from the interest in cholinergic systems that developed because of the marked changes in choline acetyltransferase in the Alzheimer brain.

The cholinergic system in the hippocampus is also involved in Alzheimer's disease. This too is a projection system with choline acetyltransferase found within nerve terminals in the hippocampus. These nerve terminals arise from nerve process projections from cell bodies located near the diagonal band of Broca, the remnant of the septal region in the human brain.

What are the behavioral functions of the cholinergic septo-hippocampal and the basocortical systems? Several years ago, Drachman and Levitt discovered that persons given a drug, scopolamine, that blocks the cholinergic receptor developed confusional states that in many ways resembled the changes seen in Alzheimer's disease, including loss of short-term memory, disorientation, the presence of intrusions, and other psychological changes. Thus, a fruitful area of investigation has been opened up for experimental psychologists, physiologists, and neurologists interested in the relationship of these newly defined brain systems to specific behavior in animals and man. The discovery of the cholinergic system deficit has also led to trials of therapy; first with the precursors, choline and lecithin, compounds that have not produced consistent improvement in memory or other functions in Alzheimer patients; but also with other drugs that act on the cholinergic system, such as physostigmine, which blocks the breakdown of acetylcholine, a drug that produces memory improvement in some patients in the early stages of Alzheimer's disease.

Recent studies have begun to delineate the specific areas of the brain affected by the Alzheimer process. Although there is atrophy of the brain in Alzheimer's disease, the degree of atrophy overlaps that seen in the course of normal aging, since nerve cell fallout is one of the consequences of the normal aging process. However, it has been found using quantitative morphometric computer-assisted techniques that within the cerebral cortex, the degree of loss of large neurons, particularly in layers 3 and 5, in the cerebral cortex is much greater than the loss of smaller neurons in the Alzheimer brain when compared to age matched controls. The pattern of involvement of neurofibillary tangles within the hippocampal formation is moderately selective, involving entorhinal cortex, CAl, and subiculum. Neurofibrillary tangles and neuritic plaques are also found in a half dozen subcortical nuclear regions in Alzheimer brains. Thus, Alzheimer's disease is a diffuse, but not random, process that attacks specific groups of nerve cells.

Etiological Factors

Genetic Factors

With these major advances in understanding of what is happening in the brain in Alzheimer's disease, there has been increased interest in searching for etiological factors. One area of major concern is the role of genetics. In a small subgroup of families, constituting perhaps only a few hundred families within the United States, Alzheimer's disease afflicts members down through many generations, apparently as a straightforward autosomal dominant inheritance. The majority of cases, however, are sporadic, with only moderate evidence of familial predisposition. Existing estimates suggest that the chance of getting Alzheimer's disease at any given age (Alzheimer's is a very sharply age-dependent disease in terms of its incidence) is increased fourfold if one has a first-degree relative, that is, a father, mother, brother, or sister with the disorder. Concordance in identical twins is variously reported to be between 40 and 60%, indicating that environmental or other nongenetic factors play a major role.

Down's Syndrome

An interesting finding is that Down's syndrome, a condition produced by the presence of a third chromosome-21, is regularly associated with pathological changes in the brain identical to Alzheimer's in individuals who live past the age of 40. These brains contain neurofibrillary tangles, neuritic plaques, and loss of choline acetyltransferase in the cortex and the hippocampus. In this regard, it should be noted that the patients studied have almost always been severely retarded Down's patients, institutionalized in various state facilities. It is difficult to determine whether or not clinical dementia occurs in this group of Down's individuals, since cognition cannot be tested in a usual fashion, but behavioral disturbances in older Down's patients frequently do occur, whereas the young Down's child is sociable and pleasant. There are however, Down's individuals with higher IQs who are in the community. It is not known whether these individuals develop dementia in their 40s or 50s, or, in fact, whether the pathologic changes observed in the institutionalized cases also occur in those in the community. However, existing evidence does suggest a very strong linkage between some kind of abnormality produced by the extra chromosome, and therefore altered genetic information and the presence of the pathologic lesions identical to those in Alzheimer's disease.

Chromosomal Abnormalities

There are further relationships in this regard. Abnormal chromosomes in some Alzheimer's patients have been reported, but the pattern of chromosomal abnormalities is not consistent from study to study. Relatives of Alzheimer probands more often have children with Down's syndrome than do those individuals in the normal population. Down's syndrome occurs more frequently in individuals whose mothers at the time of their birth were past the age of 30; a recent study of Alzheimer probands in the state of Washington have shown that their average maternal age was 31, as opposed to an average maternal age in a general population for persons of the same chronologic age at of about 23 years. This finding has not yet been replicated. Thus, there are suggestive pieces of information that require further epidemiological and cytological studies of possible chromosomal abnormalities that may increase the risk of Alzheimer's disease.

Transmissibility of Alzheimer's Dementia

Is Alzheimer's disease transmissible? The slow virus group at the National Institutes of Health has reported two instances of familial Alzheimer's disease in which a dementing illness was passed into primates by implantation of Alzheimer tissue in the brain of these primates. However, this group has not been able to replicate this finding even with additional tissues that had been stored from the same patients. Moreover, the pathology produced in these chimpanzees was that of Creutzfeld-Jakob disease rather than Alzheimer's. Thus, this finding may have been an artifact. However, the very discovery that Creutzfeld-Jakob disease, previously assumed to be a degenerative disease, was caused by a latent viral-like agent does provide a model that might be applicable to Alzheimer's disease. There are now several viral infections known to persist in the body for many years and to flare up in a neurotropic phase. Thus, Herpes Zoster, or shingles, is produced by the same virus that produced chicken pox in childhood after it has lain dominant for many years. Subacute sclerosing panencephalitis, a devastating disease often occurring in late adolescence or early adult life, is a late sequela of measles, the virus again remaining dormant for many years. Could Alzheimer's disease also be caused by a transmissable agent? This is an area where investigation almost certainly requires the availability of fresh brain tissue and where cerebral biopsy might be sought by investigators with cell culture or other systems in which to test the transmissability of an agent in Alzheimer's disease.

Environmental Factors

Another approach to etiology is to seek environmental factors that might increase the risk for or precipatate the occurrence of Alzheimer's disease. Aluminum sometimes accumulates in the Alzheimer brain, especially in relationship to the neurofibrillary tangle. Does exposure to aluminum predispose to Alzheimer? Are there abnormalities in parathyroid metabolism hormone that regulates aluminum metabolism? Do stress factors such as menopause, operations, and loss of spouse help incite the onset of the disease? These kinds of questions can best be studied epidemiologically either by use of case control studies, or prospectively in longitudinal studies of suitable populations. Such studies require the cooperation of volunteers participating in epidemiological interviews, in clinical histories and examinations, in neuropsychological examinations, blood sample programs CT scan studies, and, in some instances, lumbar puncture studies. Several such studies now have begun, and there is now preliminary data identifying specific risk factors.

These risk factors for Alzheimer's disease at present include—in addition to age and familial history—the occurrence of (up to 30 years before the onset of symptoms) significant head trauma with a period of unconsciousness.

Alzheimer's disease is malignant. It is malignant in the sense that the patient is stripped of his individuality as he or she loses all cognitive functions. It is also malignant in that life span is shortened. Patients with Alzheimer's may live from one to twenty years. In general, remaining life expectancy is reduced in half from time of onset, although duration for an individual patient cannot be predicted. An important prospective study of the malignancy of dementia was that of Nielsen, who studied the elderly population (average age in the mid-70s) on an island in Denmark in 1960 and then followed the subjects for 15 years. Mentally intact subjects survived for up to 15 years, those with severe dementia had all died within 5 years, and those with mild dementia had an intermediate life span. Improvements in medical care in 1980 may now prolong lives of severely demented individuals, but life span remains reduced.

What causes death? The proximate causes listed on death certificates are usually mundane—bronchopneumonia, myocardial infarct, and so on. But do these terminal events occur simply because of inanition, or are they related to the brain changes more specifically? Thus, we have found that community-residing Alzheimer patients have an increased degree of immunosuppression compared to age-matched normals, making them more liable to infections. An increase in immunosuppression occurs in animals after hypothalamic lesions and Alzheimer's disease often affects the hypothalamus. The Alzheimer process damaging central control of important body functions might specifically increase liability to fatal terminal events.

Summary

Thus, there has been considerable increase in our understanding of Alzheimer's disease, but we have a long way to go. Since Alzheimer's is a purely human disease, all investigations require participation of patients either for clinical evaluation, drug trials, availability of fluids such as blood and spinal fluid, and in some instances, cerebral biopsy. Availability of brains from autopsy is absolutely critical to continuing research. Among the problems that must be dealt with is the question of how one is to decide upon the risk and benefit of any particular research protocol, and how informed choice is to be obtained from the patient. Continuation of research progress is essential if we are to deal with this malignant degenerative disease, but such research must be carried out under conditions that protect the rights of the impaired patients. This is a dilemma that is clearly a most difficult one.

References

R. D. Terry and R. Katzman, "Senile Dementia of the Alzheimer's Type," *Annals of Neurology* 14, 497–506, 1983.

R. Katzman, editor, "Biological Aspects of Alzheimer's Disease," *Banbury Report* 15, New York, 1983.

Current Regulations for the Protection of Human Subjects

Charles R. McCarthy

The research community in this country has at last begun to move beyond the fundamental questions concerning the rights and welfare of human subjects involved in research. Already, the fundamental questions pertaining to persons known to be competent have been answered with sufficient clarity to provide a framework that now allows us to begin to address the more difficult questions involving persons whose capacity to provide informed consent may be limited, intermittent (in the case of Alzheimer's patients), or gradually decreasing. In order to approach the difficult questions associated with research on Alzheimer's patients, I think it is important to historically review how some of the current ethical views about the protection of the rights and welfare of human subjects developed.

In order to do so, I should like to present a skeletal historical framework that is intended to set current problems in their context and to examine some of the sociopolitical forces that have shaped this development.

The recent history of the development of ethical and legal policies for the protection of human subjects falls rather neatly into four periods of development, though of course the larger history can be traced all the way back to pre-Christian Greek and Mesopotamian cultures. My remarks here are confined to the contemporary period beginning with the revelations at Nuremberg of atrocities committed in the name of science. I have chosen to identify the periods as follows:

I. The Period of Growing Awareness: 1947–1959.
II. The Period of Policy Establishment: 1959–1966.
III. The Period of Regulatory Growth: 1966–1981.
IV. The Period of Implementation: 1981–19—.

The Period of Growing Awareness: 1947–1959

Period I includes the period from the revelations at the Nuremberg trials to the Kefauver hearings beginning in 1959. The Revelations of atrocities committed in the name of science by Nazi-controlled research investigators led to the publication of the Nuremberg Code in 1947. This code dealt almost exclusively with the dignity of competent human adults and the ethical imperative to obtain informed consent from them before involving them in research. It was immediately accepted by the world community. The Nuremberg Code was confined primarily to questions of informed consent by competent adults and therefore did not address questions of consent faced by those concerned chiefly with Alzheimer's patients, namely: How can informed consent be obtained from persons whose competence is lacking, questionable, or diminishing?

The period of growing awareness was characterized by voluntary action on the part of a small number of pioneering institutions that set out to establish mechanisms, including the first Institutional Review Boards (IRBs), for the protection of human subjects. Perhaps the most significant characteristic of this period was the general agreement with the principle that review procedures can best be carried out locally, and that governmental decision-making, if any, should be restricted to procedural requirements. This principle persists to the present time.

The Period of Policy Development: 1959–1966

The second period between 1959–1966 began and ended with writings by Dr. Henry Beecher, who put an end to the complacent view that abuses of human subjects cannot occur in this country. He presented numerous case histories in support of his position. It was during this period that the Kefauver-Harris amendments to the Food, Drug, and Cosmetic Act introduced new requirements for informed consent in drug testing. It was also in this period that several investigators created a scandal at Jewish Chronic Diseases Hospital when they injected live cancer cells into terminal cancer patients who were not given the opportunity to, or were not competent to, consent.

This period was characterized by a number of studies seeking to develop appropriate mechanisms for the protection of human subjects. Studies by the Boston University Center for Law and Medicine, policy papers developed by an NIH Committee, and recommendations by the National Health Advisory Committee all sought appropriate procedures for bringing the ethical principles that had been widely accepted since Nuremberg into some workable procedural framework.

The Period of Regulatory Growth: 1966–1981

During this period the Public Health Service (PHS) published a policy that was revised in 1966, 1967, 1969, and adopted as a Health, Education, and Welfare (HEW)–wide policy in 1971. In 1974 this policy was published in regulatory form for the first time. It was carefully examined by the National Commission for the Protection of Human Subjects of Biomedical and Behavioral Research which was created by P.L. 93–348 only six weeks after the regulations came into existence. During the four years of its existence, the National Commission issued a series of reports relating to the protection of human subjects culminating in the 1978 report on IRBs.

The HHS policies and regulations have always been characterized by three major features:

1. They require review of research by a local committee (which later came to be known as an IRB) before an investigator is permitted to carry out research with PHS (later HEW and still later Health and Human Services [HHS]) funds.
2. They require a careful risk–benefit assessment by the IRB. Research is permitted only in cases in which benefits outweigh risks. Assessment of benefits includes both potential advantages for the subjects and the value of the knowledge to be gained.
3. They require legally effective informed consent which has been described with increasing specificity as the regulations have evolved.

The Commission's IRB report was published in the *Federal Register* for public comment. In August 1979 the HEW and the FDA issued simultaneous proposals that were congruent in all matters pertaining to the structure and functions of IRBs and informed consent, and that sought to implement the recommendations of the National Commission. Altogether, the Department reviewed nearly 700 sets of comments, as well as transcripts of three public hearings on these proposals. Interest ran high and the proposals were the center of considerable discussion and controversy. In 1980 the President's Commission for the Study of Ethical Problems in Medicine and Biomedical and Behavioral Research made its own contribution to the discussions surrounding the proposed regulations.

The Period of Implementation: 1981–the Present

In January 1981, final regulations were published. They became effective on July 27, 1981. With respect to IRBs and informed consent, the new FDA and HHS rules are virtually identical. The new regulations mark the beginning of a new era in the protection of human subjects. With the publication of the new rules, discussion appears to have moved away from consideration of the substantive content of the rules and is concerned more with their implementation in a fair, efficient, and comprehensive manner.

Current efforts center on education, oversight, investigation, and reporting. Implementation is being carried out in a time when federal expenditures and personnel are being sharply cut, when federal regulations are being rolled back, and when institutional objection to federal intrusion is on the increase.

Consequently, educational efforts on the ethical dimensions of research with Alzheimer's patients appear to offer the best hope and opportunity for future development.

Let me turn my attention next to the major characteristics of the new regulations, with special emphasis on those features that tend to throw light on approaches to informed consent for persons suffering senile dementia of the Alzheimer's type.

The new regulations have four novel features:

1. They exempt specified broad categories of low-risk or risk-free research. Most of these categories fall in the areas of educational and social science research.
2. They provide for expedited review of eight specified categories of research providing that the IRB reviewers find the risks to subjects to be no more than minimal.
3. They encompass a common core of regulations shared with the FDA so that the HHS and FDA regulations are now virtually identical in all matters pertaining to the composition, responsibilities and procedures of IRBs.
4. Finally, they contain some interesting new provisions pertaining to informed consent. These provisions offer considerable discretion to investigators, providing that the IRB approves the procedures for obtaining informed consent.

Let me elaborate:

The new regulations distinguish between informed consent procedures and documentation of informed consent. Although documentation is important, the establishment of sound procedures is considered to be more important, particularly when one is dealing with patients suffering from Alzheimer's disease. Sec. 46.116 states:

> *An investigator shall seek [legally effective informed consent] only under circumstances that provide the prospective subject or the [subject's] representative sufficient opportunity to consider whether or not to participate and that minimize the possibility of coercion or undue influence. The information that is given to the subject or the [subject's] representative shall be in language understandable to the subject or the representative.*

I should like to comment here that since there may be situations where the competence of the subject is in doubt, an IRB would be acting within its authority if it required informed consent both from the subject to the extent that the subject is able to provide it, and from the subject's legally authorized representative. This would help in at least two ways: (1) If the subject is not now competent, then at least the subject will be properly represented; and (2) even if the subject is presently competent, that competence may diminish or be entirely lost. In such a case the representative may decide, on behalf of the subject, whether the subject should continue to participate in the research or not. The representative is likely to be much more effective if he or she is involved in the informed consent procedures from the beginning.

The regulations do not require an IRB to act in this way, but they encourage this kind of procedure, which would, in my judgment, provide at least one way to respect the rights of subjects, meet the requirements of the rules, and both initiate and complete research with subjects afflicted with progressive senile dementia.

In Sec. 46.111(b) dealing with criteria for IRB approval of research, the regulations state:

> *Where some or all of the subjects are likely to be vulnerable to coercion or undue influence, such as persons with acute or severe physical or mental illness [the IRB may require that] appropriate additional safeguards have been included in the study to protect the rights and welfare of these subjects.*

Such safeguards might include:

1. Appointment of a consent auditor.
2. Periodic efforts to assess the competence of the subject.
3. Including families or next-of-kin or close friends in the consent process.
4. Careful monitoring of the research.
5. Enlisting the aid of the courts in determining who can function as a legally authorized representative.

None of these procedures is appropriate in every case. All of them are appropriate in some cases. By raising these possibilities I hope to stimulate the reader to think of others and to sensitize the community of researchers who work in this field to find ways to advance research while continuing to respect the rights and welfare of the human beings who, by their participation as subjects, are partners in the research enterprise.

Note: This article was written by Dr. McCarthy in his private capacity. No official support or endorsement by the National Institutes of Health is intended or should be inferred.

Clinical Symptoms Accompanying Progressive Cognitive Decline and Alzheimer's Disease

Relationship to "Denial" and Ability to Give Informed Consent

Barry Reisberg, Betty Gordon, Martin McCarthy, and Steven H. Ferris

Introduction

Alzheimer's disease patients suffer from significant mental decline, and the capacity of these patients to provide "informed" consent can appropriately be questioned. Emotional changes accompany cognitive deterioration in the Alzheimer's patient, and further confound the patient's ability to make "reasoned" judgments about participation in research protocols. The magnitude and nature of these cognitive and emotional changes are of clear relevance for issues related to informed consent. Accordingly, in this chapter we will review the nature of these cognitive and emotional changes in mildly impaired, and severely impaired Alzheimer's patients, and explore the relevance of these changes for issues related to informed consent.

We can divide the syndrome of age-associated cognitive decline, and progressive Alzheimer's disease into three clinical phases. [1,2] The condition is sufficiently common that each of the phases should be readily recognizable to geriatricians and nongeriatricians alike, both from experiences with patients as well as from experiences with one's own family members and acquaintances.

The earliest, very mild phase may be termed the Forgetfulness Phase. In this phase, the cognitive deficit is primarily subjective. The individual, and his or her spouse, notice a tendency to forget where objects have been placed. Also, the individual in this phase is aware of increased difficulty in recalling the names of persons and places. The Forgetfulness Phase individual may also have more difficulty with appointments and finds a need to write things down more frequently in order to remember them. These symptoms can frequently be objectified utilizing psychometric assessments. On these assessments, Forgetfulness Phase persons may display a relative deficit for their age in associative memory tasks. However, in general, the symptoms do not interfere significantly with employment or daily performance. The symptoms are accompanied by an increase in anxiety that is probably adaptive and that in most cases does not require treatment.

This condition is followed by one of definite impairment in which mild to moderate deficit becomes clinically evident. This may be termed the Confusional Phase. The cognitive deficit is particularly notable for memory of recent events. Deficits in past memory are less evident, but nevertheless are present as well. Concentration ability is also frequently affected. Vocabulary is largely spared; however, the individual may experience difficulty recalling appropriate words. There is often little in the way of symptomatology apart from the cognitive deficit.

The moderately severe to very severe Dementia Phase may be defined as beginning at the point at which individuals can no longer survive if left on their own. Early in this phase, individuals require assistance in such basic activities as dressing and shopping. They are no longer able to select the proper clothing, but remain capable of dressing themselves. They may be able to travel to the corner grocery and return, but lose track of their purchases. As this phase progresses, all ability to carry out the activities of daily living is lost. Individuals lose the ability not only to dress, but to eat and toilet themselves. Memory suffers to the extent that they can no longer name the spouse upon whom they are entirely dependent for survival and, ultimately, Dementia Phase patients forget their own names.

These broad phases of dementia can be further subdivided into several stages of cognitive decline, from the stage of no deficit to severe dementia. The Global Deterioration Scale (GDS) for Age-Associated Cognitive Decline and Alzheimer's Disease is based upon these seven clinically identifiable stages[3] (*see* Table 1). Previous investigations have shown strong relationships between assessment on this instrument and independent behavioral,[4] neuroradiologic,[5,6] neurometabolic,[7] electrophysiologic,[8] and neuroimmunologic[9] measures in outpatients with cognitive decline consistent with normal aging and/or with Alzheimer's disease.

Accompanying this process of progressive decline in cognitive functioning are two psychological processes that have clear relevance with respect to the ability of patients to render knowledgeable, informed consent. One is a progressive decrease in insight and knowledge with increasing severity of the illness process. The second is the extent to which the patient denies illness and, hence, fails to recognize the relevance of attempts to treat the illness process.

The loss of one's intellectual and general thinking capacities is a terrible tragedy, too painful for conscious contemplation. As with any devastating illness or loss, the psychological mechanism of defense termed "denial" operates to prevent full conscious contemplation of a loss that would be emotionally overwhelming. Psychiatrists define "denial" as "a defense mechanism, operating unconsciously, used to resolve emotional conflict and allay anxiety by disavowing thoughts, feelings, wishes, needs, or external reality factors that are consciously intolerable."[10] "Denial is usually betrayed by the obvious disparity between the patient's condition and how he reports it. Many such patients smilingly insist that all is well or that a symptom does not exist."[11]

Although denial has been reported to accompany a very broad spectrum of physical and emotional maladies, the extent to which this mechanism operates as a concomitant of cognitive decline in normal aging and in Alzheimer's disease has only recently been systematically studied. The dimensions of this psychological process are of importance to clinicians and investigators for a variety of reasons. One reason is that the validity of self-assessments of cognitive and functional status has clear relevance for issues related to informed consent. An individual

Table 1. Global Deterioration Scale (GDS) for Age-Associated Cognitive Decline and Alzheimer's Disease

GDS stage	Clinical phase	Clinical characteristics	Psychometric concomitants
1 No cognitive decline	Normal	No subjective complaints of memory deficit. No memory deficit evident on clinical interview.	Average or above average performance for age and WAIS vocabulary score on 3 of 5 Guild memory subtests.
2 Very mild cognitive decline	Forgetfulness	Subjective complaints of memory deficit, most frequently in following areas: (a) forgetting where one has placed familiar objects; (b) forgetting names one formerly knew well. No objective evidence of memory deficit on clinical interview. No objective deficits in employment or social situations. Appropriate concern with respect to symptomatology.	Below average performance for age and WAIS vocabulary score on 3 of 5 Guild subtests.
3 Mild cognitive decline	Early Confusional	Earliest clear-cut deficits. Manifestations in more than one of the following areas: (a) patient may have gotten lost when traveling to an unfamiliar location; (b) co-workers become aware of patient's relatively poor performance; (c) word and name finding deficit become evident to intimates; (d) patient may read a passage or a book and retain relatively little material; (e) patient may demonstrate decreased facility in remembering names upon introduction to new people; (f) patient may have lost or misplaced an object of value; (g) concentration deficit may be evident on clinical testing.	One standard deviation or greater below average performance for age and WAIS vocabulary score on three of five Guild memory subtests. Often no errors on the Mental Status Questionnaire (MSQ).

		Objective evidence of memory deficit obtained only with an intensive interview conducted by a trained geriatric psychiatrist. Decreased performance in demanding employment and social settings. Denial begins to become manifest in patient. Mild to moderate anxiety accompanies symptoms.	
4 Moderate cognitive decline	Late Confusional	Clear-cut deficit on careful clinical interview. Deficit manifest in following areas; (a) decreased knowledge of current and recent events; (b) may exhibit some deficit in memory of one's personal history; (c) concentration deficit elicited on serial substractions; (d) decreased ability to travel, handle finances, etc. Frequently no deficit in following areas: (a) orientation to time and person; (b) recognition of familiar persons and faces; (c) ability to travel to familiar locations. Inability to perform complex tasks. Denial is dominant defense mechanism. Flattening of affect and withdrawal from challenging situations occur.	Frequently mistakes on 3 or more items on MSQ.
5 Moderately severe decline	Early dementia	Patient can no longer survive without some assistance. Patient is unable during interview to recall a major relevant aspect of their current lives: e.g., their address or telephone number of many years, the names of close members of their family (such as grandchildren), the name of the high school or college from which they graduated. Frequently some disorientation to time (date, day of week, season, etc.) or to place. An educated person may have difficulty counting back from 40 by 4s or from 20 by 2s.	Deficits evident on brief MSQ assessment.

(continued on next page)

Table 1. Continued

GDS stage	Clinical phase	Clinical characteristics	Psychometric concomitants
		Persons at this stage retain knowledge of many major facts regarding themselves and others. They invariably know their own names and generally know their spouses and children's names. They require no assistance with toileting or eating, but may have some difficulty choosing the proper clothing to wear and may occasionally clothe themselves improperly (e.g., put shoes on the wrong feet, etc.).	
6 Severe cognitive decline	Middle dementia	May occasionally forget the name of the spouse upon whom they are entirely dependent for survival. Will be largely unaware of all recent events and experiences in their lives. Retain some knowledge of their past lives but this is very sketchy. Generally unaware of their surroundings, the year, the season, etc. May have difficulty counting from 10, both backward and sometimes, forward. Will require some assistance with activities of daily living, e.g., may become incontinent, will require travel assistance, but occasionally will display ability to travel to familiar locations. Diurnal rhythm frequently disturbed. Almost always recall their own name. Frequently continue to be able to distinguish familiar from unfamiliar persons in their environment.	5–10 errors on MSQ.

		Personality and emotional changes occur. These are quite variable and include: (a) delusional behavior, e.g., patients may accuse their spouse of being an impostor; may talk to imaginary figures in the environment, or to their own reflection in the mirror; (b) obsessive symptoms, e.g., person may continually repeat simply cleaning activities; (c) anxiety symptoms, agitation, and even previously nonexistent violent behavior may occur; (d) cognitive abulia, i.e., loss of willpower because an individual cannot carry a thought long enough to determine a purposeful course of action.
7 Very severe cognitive decline	Late dementia	All verbal abilities are lost. Frequently there is no speech at all—only grunting. Incontinent of urine; requires assistance toileting and feeding. Lose basic psychomotor skills, e.g., ability to walk. The brain appears to no longer be able to tell the body what to do.
		Generalized and cortical neurologic signs and symptoms are frequently present.

who is unwilling or unable psychologically to accept the existence of a symptom or illness may be unwilling or psychologically unable to consent to treatment of that condition. If the lack of recognition of the illness is a product of decreased insight, itself an invariable symptom of the illness process, that may have somewhat different implications for issues related to informed consent than if the lack of recognition is the product of denial or other so-called "psychological mechanisms of defense." For example, denial may cause patients actively to avoid treatment for their illness at a stage when they are otherwise capable of understanding what is happening to them. Also denial may, in general, increase patients' resistance to participation in research that confronts them with the reality of their illness. This increased resistance may cause otherwise cooperative, if less aware and insightful, patients to avoid participation in research.

In a recent study conducted in our laboratory, we attempted to outline the manifestations of denial with progressive cognitive decline in normal aging and in Alzheimer's disease. In this study, 35 community-residing couples consisting of a subject and a spouse were interviewed. The subjects were 60 to 85 years of age and consisted of controls ($N = 10$), subjects with a primary diagnosis of age-associated cognitive decline consistent with senescent forgetfulness ($N = 5$), and subjects with Alzheimer's disease ($N = 25$). Subjects with a history of acute or chronic illnesses of sufficient severity to interfere with cognition were excluded from participation. Exclusion criteria included history of psychiatric hospitalization or significant affective disorder, alcohol or other drug abuse, an acute or chronic illness of sufficient severity to interfere with cognition, and a history of stroke or symptomatology indicative of multi-infarct dementia. Subjects were interviewed and asked a series of questions with respect to their own functioning, and an identical series of questions with respect to their spouses' functioning. Spouses were interviewed separately and asked a series of questions with regard to their own functioning and an identical series of questions with regard to the patients' functioning. Questionnaires utilized for both patients and spouses were the same. The results for a few representative queries will be reviewed briefly because they illustrate the process of denial in these patients and its relationship to progressive lack of insight. We will then discuss the precise relevance of these findings for issues relating to informed consent.

When asked about memory problems (Fig. 1), very mildly impaired Forgetfulness Phase patients (GDS 2) rated their problem as being somewhat worse than did those patients with no impairment (GDS 1). Similarly, the mildly impaired, Early Confusional Phase patients (GDS 3) rated their problems as being considerably worse than the Forgetfulness Phase patients. However, after this Early Confusional Phase, patients with

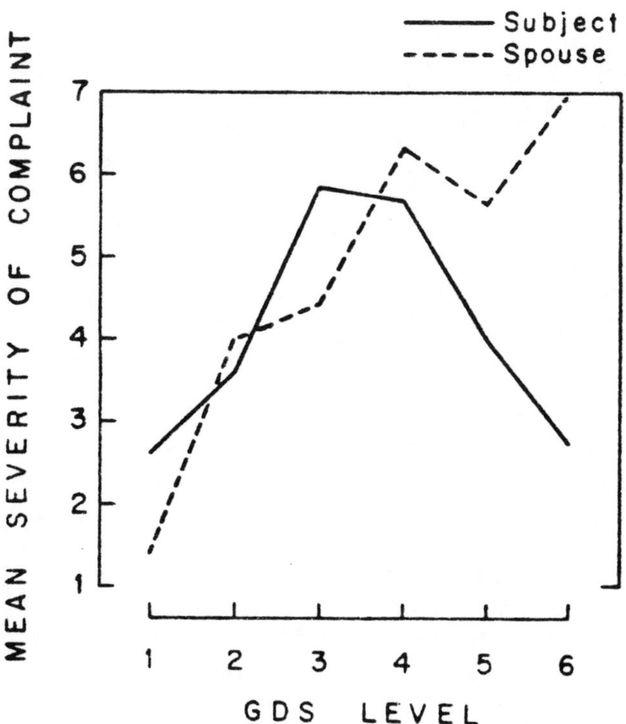

Fig. 1. Questions in reference to the patient's status: Category 1: Memory functioning. Query 1: What kinds of problems do you (does your spouse) have with memory?

progressively increased levels of objectively rated impairment assessed the magnitude of their memory problems as progressively less severe. Spouse assessments of the magnitude of the patients' memory problems, in contrast, continued to increase, more or less regularly, with increments in objectively assessed deficit.

With respect to difficulty in recalling recent events, we observe a pattern very similar and equally dramatic to that which we have described in response to the general question with respect to memory problems in Fig. 1. Patient assessments of difficulty in recalling recent events peaked in the Late Confusional Phase (GDS 4) and thereafter, more severely impaired patients assessed their difficulties in recalling recent events as virtually nonexistent. Once again, spouse assessments of difficulties in this area tended to rise steadily with increments in objectively assessed impairment in the patient.

When queried with respect to emotional problems experienced as a result of memory difficulties, patients' assessments of the emotional concomitants paralleled almost precisely the extent to which they had rated the severity of their memory problem in Query 1. Spouse assessments, however, now appeared to mirror the patients' own assessments of emotional difficulties. The spouse appeared to appreciate the fact that psychological mechanisms were acting to prevent the patient from experiencing emotional upset, although they recognized that the patient's memory problem continued to worsen. However, the patients with moderate (GDS 4) to severe (GDS 6) memory impairment continued to rate the magnitude of their emotional difficulties considerably less than their spouses. This would seem to indicate either that these patients were denying emotional problems as well as memory problems, or that the spouses failed to appreciate the extent to which psychological mechanisms of defense were acting to prevent the patients' conscious experience of emotional upset.

In contrast to our findings with respect to queries relating to memory and cognitive functioning, we find that patients' and spouses' assessments of their ability to communicate with each other follow each other closely throughout the course of the illness, indicating insight on the part of the patient in this area. However, we also note a slight increase in anxiety with respect to communication within the marital relationship on the part of the Forgetfulness Phase patient. Patients and their spouses were also not far apart in their assessments of the patients' satisfaction with their mutual sexual relationships.

Although it is difficult to separate, in a definitive fashion, the lack of insight associated with the illness process from the psychological defense mechanism that we term denial, we attempted to further separate these factors by asking both the patient and the spouse an identical series of questions, this time relating to the spouse's condition. Regardless of how they had assessed the magnitude of their own memory problem, patients' recognized that their spouses did not have significant problems in this area. Similarly, patients and spouses were in complete agreement with respect to the absence of "a sense of confusion or loss of orientation," or difficulties with recent memory on the part of the spouse.

Hence, patients appeared to continue to display insight with respect to their spouses' status throughout the course of the illness studied. Their "lack of insight" appears to have been selective for processes affecting themselves, and thus appears to have been the product of a defense mechanism, specifically, denial.

Relevance of Findings for Issues Related to Informed Consent

The results indicated that the earliest symptoms of cognitive decline (the Forgetfulness Phase) are fully recognized both by the patient and by those with whom they are in most intimate contact—their spouses. In a sense, the observational powers of the spouse are validated by the remarkable concordance of both patients and their spouses with respect to the onset and severity of these very subtle early cognitive symptoms. Emotionally, these early symptoms evoke a sense of alarm on the part of both patients and their spouses. Both recognize increased emotional difficulties that noticeably affect their family relationship.

Both patients and their spouses become somewhat more irritable as a result of these symptoms. The spouses, in particular, are somewhat ashamed of the patients' forgetfulness; however, neither the patient nor the spouse feels at all helpless at this early stage. Interestingly, at this early stage the patient becomes not only acutely aware of a personal cognitive problem, but also acutely, or perhaps hyperacutely, sensitive to slight cognitive problems in the spouse.

These processes have direct effects on the willingness of patients with these early Forgetfulness Phase symptoms to participate in research. We find that not only are these patients and their spouses very aware of their symptoms, but they are anxious to do whatever they can to alleviate them. As we have seen, the symptoms are met not with a sense of helplessness, but of appropriate concern. Our experience with thousands of patients over the course of nearly a decade of research in this area has been that the Forgetfulness Phase subjects are the most available, and perhaps the most willing, to participate in research that might ameliorate their symptomatology. Since their insight, judgment, and general cognitive abilities are not seriously impaired at this stage, these patients are fully capable of weighing comprehensible research consent forms in terms of their benefits and risks, and of giving knowledgeable and informed consent. Their increased anxiety and shame, as a result of their symptomatology, is likely to increase their readiness to participate in research, but is not likely to significantly affect intellectual judgments.

In general, the patients' awareness of their problem tends to peak in the Confusional Phase (mild to moderate impairment). Spouses' awareness of memory problems in the patient tend not to differ markedly from the patients' assessments at this phase. Patients and their spouses continue to experience some emotional problems as a result of the patients' memory difficulties. However, increased irritability and shame are transient phenomena that the patient is able to suppress at this phase. A sense of helplessness on the part of both the patient and the spouse also develops for the first time in this phase. Confusional Phase patients and their spouses appear to be capable of adjusting socially to the patients' memory problems and somewhat isolating the cognitive symptomatology in terms of its marital and social manifestations. Denial of specific cognitive problems does occur in the Confusional Phase, however. Specifically, patients, but not their spouses, are unwilling to accept that they might be less capable of carrying out their basic activities of daily living.

Several of the above factors are relevant with respect to the ability of the Confusional Phase subject to give informed consent. Since the patients recognize their memory deficit and can somewhat adequately assess its seriousness, they should be capable of making informed decisions regarding useful or experimental treatments. However, the sense of helplessness on the part of both the patient and his or her spouse with respect to the patient's memory problem, which replaces the former irritability and sense of shame, undoubtedly affects decisions to enter into research projects. In general we find that patients at this phase do tend to be willing to participate in research with respect to their illness, but are less eager participants than the Forgetfulness Phase subjects. Also, their sense of helplessness makes patients at this phase somewhat less likely to volunteer for participation.

Another issue is whether the patient's illness itself (in the Confusional Phase) with the clinically evident decrease in cognitive functioning, and necessarily impaired insight and judgment, itself makes informed consent for participation in research impossible. The answer appears to be "no." Since patients do appear to be capable of "reasonable" assessments with respect to their cognitive functioning, they are probably capable of reasonably assessing their participation in projects to ameliorate the deficit. Undoubtedly, clarity of presentation of the research protocol and research design become particularly important in informing the Confusional Phase subject.

The one exception to the above rule is with respect to programs designed to enhance the ability of the patient to carry out the complex activities of daily living. Since patients, but not their spouses, deny deficits in this area, they might be less likely than their spouses to accept participation in cognitive training or other programs designed to increase their functional capacity. Indeed, in our own research, we have found patients reluctant to participate in such cognitive training programs. However, the reasons for the patients' reluctance are multiple, and include such elements as the frustration engendered by cognitive and functional exercises, as well as denial of deficit. Since the patients' active and enthusiastic participation in such activities is absolutely essential for their success, the question of participation with anything other than a patient's fully informed consent is probably moot at this phase.

A final issue in the Confusional Phase is whether it is necessary to obtain the informed consent of the spouse as well as that of the patient. The research just described has indicated that patients at this phase are capable of insight with respect to their cognitive deficit and emotional status. It should be recalled that in the Confusional Phase, many patients continue to be able to function in nondemanding job settings. Patients at this phase are always legally competent as well, in the sense of understanding the nature and extent of their possessions. Hence, requiring the consent of the spouse for participation in relevant research protocols, in addition to that of the patient, could constitute denial of legal and social status and rights that the patient continues to possess. However, investigators undoubtedly have a right to set their own criteria and standards for a patient's entry into voluntary research endeavors. By definition, patients in the confusional phase suffer from decreased cognitive capacity. Apart from denial, this decreased cognitive capacity is certainly accompanied by decreased insight and judgment. Hence, it is probably both proper and desirable for investigators to obtain informed consent from spouses as well as patients in this phase of the illness process. In general, the rule that we have followed is to require the spouse's consent as well as the patient's in those cases where the spouse finds it necessary or desirable to accompany the patient to the clinic.

In the Dementia Phase, patients develop a profound denial of cognitive and emotional deficit. The denial appears to occur in precisely those areas of cognition and emotional functioning that are most severely affected. For example, the denial appears to be somewhat less marked when patients are asked relatively oblique questions, such as "do you feel a sense of confusion or loss of orientation?" in comparison to that exhibited in response to direct questions such as regarding problems with memory. Despite the profound denial, even in the Early and Mid-Dementia Phases, patients do appear to display insight with respect to the functioning of their spouses in cognitive and other areas.

In a practical sense, denial affects participation in research in a variety of ways. Although such patients are probably in a minority, some patients in the Dementia Phase find any evaluation of their memory that forces them to begin to confront their deficit as too painful an experience for voluntary participation. Some such patients refuse to see physicians in general, and physicians or other professionals who will be evaluating their cognitive status in particular. Other such patients who are brought in for an evaluation become acutely anxious. They may develop an actual anxiety attack or exhibit conversion or dissociative symptomatology (i.e., in lay terminology, "hysterical behavior"). For example, one woman responded to all questions that were put to her by panting and grunting. Other patients simply refuse any evaluation of their memories and literally run out of the office or testing room. Agitation, of course, is a common occurrence, particularly in Mid-Dementia Phase patients.[12] This agitation is a result of a variety of interacting processes including changes in brain chemistry in general, and brain neurotransmitter changes in particular, as well as the result of cognitive and psychological processes that make a formerly benign environment suddenly profoundly threatening. Nevertheless, even patients who are not profoundly agitated prior to cognitive assessments often become agitated in the course of such evaluations. This increased agitation in many instances appears to be a direct result of patients being confronted with knowledge of their profound intellectual losses.

Many Dementia Phase patients, although they display marked denial symptomatology, do not exhibit the extreme symptoms described above. For example, a typical patient when asked "who is the President of the United States?" will, not knowing the answer, simply respond "I don't follow politics very closely."

With respect to participation in research protocols, it is probably unwise and counterproductive to include patients from whom minimum tacit cooperation has not been obtained. All of our protocols at the Geriatric Study and Treatment Program include the subject exclusion criterion of "hostility or refusal to cooperate." Hence, ethical issues need only be raised for those patients in the unlikely event that a study specifically required uncooperative patients.

For the majority of Dementia Phase subjects, tacit and explicit cooperation with research protocols and study designs is obtainable. In the case of all Dementia Phase subjects, we follow the rule of always getting permission and informed consent from the spouse as well as from the patient for participation. We follow this rule because, by definition, Dementia Phase subjects are no longer as capable of caring for themselves as formerly. It should be noted that we obtain informed consent from the spouse, regardless of the legal status of the spouse as a guardian for the patient. We believe that the insight that patients continue to evidence with respect to less- and nonthreatening cognitive and emotional areas demonstrates this ability to comprehend their participation in a research project that has been properly explained to them. In a pragmatic sense, we find that although Dementia Phase patients are not willing to confront their cognitive and emotional difficulties directly, participation in programs designed to further physicians' and scientists' understanding and treatment of cognitive and emotional problems associated with aging is sufficiently indirect and nonthreatening for the majority of patients such that both cooperation and informed consent are obtainable.

One other pragmatic concern with respect to informed consent documents is the extent to which they should discuss specific diagnoses, such as Alzheimer's disease, and issues related to the diagnoses, such as prognosis and treatment. We believe that discussions between the physician, the patient, and family members or caregivers of the patient should occur at the time at which a diagnosis is arrived at. In all instances, patients should be diagnosed prior to being placed in research procedures or protocols. Furthermore, the procedures and purposes of the specific research project or protocol should be explained to the patient in detail prior to their being given a consent form. Hence, the consent form should contain only information that has been previously discussed with the patient and their accompanying family members, guardian, or caregiver. In the case of patients whose reading comprehension is impaired (patients in the Late Confusional or Dementia phases), the document should be read aloud in the presence of both the patient and other responsible persons. The consent form should contain all of the following:

1. The purpose of the study.
2. A description of the study.
3. A statement regarding participation in the study.
4. A statement of possible benefits and risks.
5. A statement regarding discomforts.
6. A statement regarding alternative therapies or procedures.
7. A statement regarding confidentiality.
8. A statement regarding research-related injury.

An example of such a consent form can be found in Table 2.

Table 2 SAMPLE CONSENT FORM

CONSENT FORM: INVESTIGATIONAL DRUG TREATMENT STUDY

Purpose of the Study

You are volunteering to participate in a research project to determine whether this investigational drug is effective in the treatment of Primary Degenerative Dementia (Alzheimer's Disease). This is a new compound which is similar to both antidepressants and stimulants, and which seems to improve memory in animals. It is not a marketed drug in the United States, and its use for the treatment of dementia is not approved. Preliminary studies have shown this compound to be safe for use by elderly patients. This study will provide further evidence for the effectiveness and safety of this drug.

Description of Study

All patients will receive the drug in dosages ranging from 1 mg, twice a day to 10 mg, twice a day. The effects of this treatment will be evaluated. The study will require your participation for 16 weeks. You will be asked to come to the clinic once a week for 6 weeks and then once every 2 weeks for 10 weeks. At the beginning, you will receive a thorough medical and neurological examination, including chest X-ray, electrocardiogram (EKG), and the taking of blood (90 cc or 2.7 oz), and urine for laboratory evaluations. You will also receive a CT scan, a special test that takes X-ray pictures of the brain. At your weekly or biweekly visits, changes in your condition will be evaluated by interview, psychological tests, and rating scales, and periodically, by additional medical and laboratory examinations.

Participation in the Study

Your participation in the study is voluntary. You may refuse to participate, and you are also free to withdraw from the study at any time. These actions will not prejudice your further treatment or participation in our program. However, if you withdraw from this study you will be asked to have a final examination to evaluate the effects of your treatment. Your participation in the study may also be ended if the physician decides that this is in your best interest. The investigational drug will not be available to you at the end of the study, but alternative treatment may be provided.

Benefits and Risks

Your condition may improve as a result of your participation in this study. However, since this drug for the treatment of dementia is still in the testing stages, it might not be better than other treatments that may be available. Although there are currently no known risks associated with the investigational drug, some undesirable side effects may occur. These may include . During the study the doctor will question you carefully about possible side effects and take steps to minimize them.

Discomforts

Punctures of a vein will be done to obtain blood samples. These will cause a pinprick sensation where the needle is inserted.

Alternative Therapies

You will be asked to avoid any other treatment for your dementia while you are in this study. While there is no generally recognized adequate treatment for your illness, Hydergine is sometimes used to provide relief of symptoms. Hydergine is believed to improve blood circulation in the brain.

Costs

You will not incur any costs as a participant in this study.

Confidentiality

Your identity in this study will be treated as confidential. However, in order to meet obligations of federal laws, records identifying you may be inspected by representatives of the sponsor of the study and/or representatives of the Food and Drug Administration. By signing this form, you consent to such inspection and disclosure.

Research-Related Injury

The Medical Center will provide essential medical care for any physical injury resulting from participation in this research project. Neither financial compensation nor long-term medical treatment for such injuries will be provided.

Information concerning your rights as a research subject or on the availability of treatment for physical injury resulting from participation in the research project may be obtained from , M.D., telephone or from the Medical Center Office of Grants Administration and Institutional Studies, telephone . Dr. will also be available at all times during the course of the study to answer any questions which may arise.

I voluntarily consent to participate in the study described above.

______________________________	____________
Patient	Date
______________________________	____________
Relative*	Date
______________________________	____________
Witness	Date
______________________________	____________
INVESTIGATOR	Date

*Please note that if the patient has been declared legally incompetent, then the kinsman cosigner must be properly qualified to authorize the patient's participation.

Conclusion

In conclusion, we find that the ability of patients to recognize cognitive and emotional deficits and to render informed consent for their participation in research protocols is profoundly affected by the magnitude of their cognitive deficit. The degree of cognitive deficit in turn results in changes not only in the patients' understanding, but also in emotional and psychological changes that in part compel a patient to deny their deficit even as it becomes more profound. These processes have both practical and theoretical implications for obtaining patients' consent for participation in research protocols.

Practically, denial of the illness will lead many severely impaired patients to resist participation in research studies that force them to confront their memory deficits. Pragmatically, such patients should probably be excluded from most research programs. Another pragmatic approach is to obtain consent from spouses or other legal guardians, as well as patients, wherever patients are diagnosed as having suffered significant cognitive impairment. One pragmatic definition of "significant cognitive impairment" is if the spouse, caregiver, or legal guardian is required to accompany the patient to the research setting. In addition to obtaining consent from responsible family members, guardians, and/or caregivers, all consent documents should be explained in detail to both the patient and the responsible person. In the case of patients in the Late Confusional phase, or of those with more severe impairment, the document should be read aloud by the physician or investigator to the patient in the presence of the responsible persons. Naturally, any and all questions should be answered at that time. Following these procedures we find that the majority of patients, including those with severe impairment, although they may not be willing to admit verbally to suffering from a severe loss of memory and intellectual ability, tacitly recognize deficits. This tacit recognition of deficit permits the investigator to obtain explicit consent for participation in research protocols from the majority of severely impaired patients, as well as from mildly and moderately impaired patients.

Notes and References

[1]B. Reisberg, *Brain Failure: An Introduction to Current Concepts of Senility* (New York: The Free Press/Macmillan, 1981), pp. 81–122.

[2]B. Reisberg and S. H. Ferris, "Diagnosis and Assessment of the Older Patient," *Hospital & Community Psychiatry*, 33, 1982, 104–110.

[3]B. Reisberg, S. H. Ferris, M. J. de Leon, and T. Crook, "The Global Deterioration Scale (GDS): An instrument for the assessment of Primary Degenerative Dementia (PDD)," *American Journal of Psychiatry*, 139, 1982, 1136–1139.

[4]B. Reisberg, S. H. Ferris, M. K. Schneck, M. J. de Leon, T. Crook, and S. Gershon, "The relationship between psychiatric assessments and cognitive test measures in mild to moderately cognitively impaired elderly," *Psychopharm Bulletin*, 17, 1981, 99–101.

[5]M. J. de Leon, S. H. Ferris, I. Blau, A. E. George, B. Reisberg, I. I. Kricheff, and S. Gershon, "Correlations between computerized tomographic changes and behavioral deficits in senile dementia," *Lancet*, ii, 1979, 859.

[6]M. J. de Leon, S. H. Ferris, A. E. George, B. Reisberg, I. I. Kricheff, and S. Gershon, "Computed tomography evaluations of brain-behavior relationships in senile dementia of the Alzheimer's type," *Neurobiology of Aging*, 1981, 69–79.

[7]S. H. Ferris, M. J. de Leon, A. P. Wolf, T. Farkas, D. R. Christman, B. Reisberg, J. S. Fowler, R. MacGregor, A. Goldman, A. E. George, and S. Rampal, "Positron emission tomography in the study of aging and senile dementia," *Neurobiology of Aging*, 1, 1980, 127–131.

[8]B. Reisberg, S. H. Ferris, H. Ahn, and E. R. John, "Electrophysiologic asymmetry and brain change in primary degenerative dementia," *Soc. Neurosci.* (Abstracts), 7, 1981, 241.

[9]K. Nandy, B. Reisberg, S. H. Ferris, and M. J. de Leon, "Brain reactive antibodies and progressive cognitive decline in the aged," *Journal of American Aging Association* (abstract), 4, 1981, 145.

[10]*A Psychiatric Glossary*, 3rd Edition (Washington, DC: American Psychiatric Association, 1969), p. 28.

[11]M. J. Martin, "Psychiatry and Other Specialties," in A. M. Freedman, H. I. Kaplan, and B. J. Sadock, eds., *Comprehensive Textbook of Psychiatry, II* (Baltimore: Williams and Williams Co., 1975), p. 1740.

[12]B. Reisberg, "The office management of primary degenerative dementia," *Psychiatric Annals*, 12(6), 1982, 631–637.

The Physician–Researcher

Role Conflicts

Robert J. Levine

What is the role of the physician? What is the role of the researcher? Are these roles inherently in conflict with each other? If so, are they so much in conflict that when one professional attempts to play both roles simultaneously we should impose special procedural protections for the rights and welfare of the patient–subjects? Or is the conflict so threatening and so incorrigible that we should forbid any professional to play both roles simultaneously?

Professional Roles

Let us begin with a deliberately simplistic examination of the roles of physician and researcher. They are, respectively, to practice medicine or to conduct research. The following definitions of these two enterprises are compatible with those adopted by the National Commission for the Protection of Human Subjects of Biomedical and Behavioral Research (the National Commission):

> *The term "research" refers to a class of activities designed to develop or contribute to generalizable knowledge. By generalizable knowledge is meant theories, principles or relationships (or the accumulation of data on which they may be based), that can be corroborated by accepted scientific observation and inference.*
>
> *The "practice" of medicine or behavioral therapy refers to a class of activities designed solely to enhance the well-being of an individual*

> *patient or client. The purpose of medical or behavioral practice is to provide diagnosis, preventive treatment or therapy.*[1]

According to these definitions, then, when a physician is practicing medicine, he or she is performing activities that are designed solely to enhance the well-being of an individual patient. Superficially, at any rate, it appears that he or she has no competing or conflicting interests. This is not necessarily the case. The physician who is not doing research may have various sorts of conflicting interests. For example, in considering whether to recommend surgery or various diagnostic tests, the judgment of some physicians may be influenced by the fact that positive recommendations yield greater financial rewards for the physician. Thus, particularly when other considerations do not clearly indicate the making of a positive or negative recommendation, the financial interests of the physician may, in some cases, "tip the balance" toward the positive recommendation.

How do we safeguard the interests of the patient against such competing interests? In my view, the most powerful safeguard is and ought to be a reliance on the professional responsibility of the physician. By this I mean a reliance on the integrity of each individual physician as well as a reliance on the social pressure that can be brought to bear by his or her colleagues. There is, of course, a second line of defense. This is reflected in various laws and institutional policies requiring second opinions for certain sorts of elective surgery and in PSRO requirements for continual review of ongoing activities.

Rather than attempt to develop an exhaustive or extensive list of potential conflicts of interest in the practice of medicine, let us just agree that there are some. What is different in research is that, by definition, there is invariably something being done that is designed to benefit someone or ones other than the subject. Thus, when an professional assumes the dual role of physician–researcher in relation to a patient–subject, there is an inherent conflict. Although this conflict is not necessarily different in kind from those present in the practice of medicine,[2] it differs in that in medical practice, the usual presumption should be that there is no important conflict of interest. In research, on the other hand, there is no presumption; there is invariably the knowledge that there is a conflict.

Some Considerations of the Conflict

Most of the published commentary on this conflict centers on the issue of informed consent. Who should negotiate informed consent with the patient–subject? Should it be the personal physician, the researcher, or the physician–researcher? Should there be in some cases another agent involved, such as a consent auditor or an advocate? Although federal regulations are silent on this matter, the leading ethical codes are not. According to the Nuremberg Code:

> *The duty and responsibility for ascertaining the quality of the consent rests upon each individual who initiates, directs or engages in the experiment. It is a personal duty and responsibility which may not be delegated to another with impunity.*

The Nuremberg Code, of course, is not concerned with such complicated roles as physician–researcher or patient–subject. It is designed to provide guidance to researchers who are using nothing but nontherapeutic procedures. Though they might also be physicians, when they are acting according to the guidance provided by Nuremberg, they are performing exclusively in the role of researcher.

Many commentators on the practice of medicine have expressed concern about the imbalance of power between the physician and the patient. Many of these commentators have drawn upon Talcott Parsons' perspectives on the social role of "sick person," the privileges and responsibilities of the role, and the dependency of the sick person upon the physician to "legitimate" that role.[3] Thus, when a "sick person" is invited to perform also in the role of subject, there is great concern about the potentialities for exploitation of this imbalance of power.

There has been considerable debate about whether a physician who is involved in a physician–patient relationship can negotiate fairly for informed consent with the patient to become a subject.[4] Spiro, for example, asserts that a physician having a close relationship with a patient can usually persuade that patient to do almost anything.[5] Unlike most commentators, because Spiro emphasizes the importance of the closeness of the relationship, he feels the problem is greater in private practice than it is with ward or clinic patients. Henry Beecher[6] reviewed the literature on this subject; in his conclusion he suggests that consent might not be either the only or the most important issue[7]

> *An even greater safeguard for the patient than consent is the presence of an informed, able, conscientious, compassionate, responsible investigator, for it is recognized that patients can, when imperfectly informed, be induced to agree, unwisely, to many things. . . .*
>
> *A considerable safeguard is to be found in the practice of having at least two physicians involved. . . . First there is the physician concerned with the care of the patient, his first interest is the patient's welfare; and second, the physician–scientist whose interest is the sound conduct of the investigation. Perhaps too often a single individual attempts to encompass both roles.*

Beecher was not clear about which of these two physicians he would have negotiate informed consent. The Declaration of Helsinki requires the following (Principle I.10):

> *When obtaining informed consent for the research project, the doctor should be particularly cautious if the subject is in a dependent relationship to him or her or may consent under duress. In that case the informed consent should be obtained by a doctor who is not engaged in the investigation and who is completely independent of the official relationship.*

In its report on IRBs, the National Commission suggests in its commentary under Recommendation 3D that the IRB should be aware of the advantages and disadvantages (for patient–subjects) of having one individual perform the dual role of the physician–researcher. At its discretion, the IRB may require a "neutral person" not otherwise associated with the research, or the investigator may be present when consent is sought or to observe the conduct of the research. This "neutral person" may be assigned to play a role in informing subjects of their rights and of the details of protocols, assuring that there is continuing willingness to participate, determining the advisability of continued participation, receiving complaints from subjects, and bringing grievances to the attention of the IRB. Federal regulations developed in response to the recommendations of the National Commission do not reflect these considerations.

The National Commission was more explicit on this point in its Report on those institutionalized as mentally infirm; of the various Reports of the National Commission, this Report seems most particularly relevant to the problems we are discussing at this Conference. Recommendation 1H states that the IRB must determine that:

> *Adequate provisions are made to assure that no prospective subject will be approached to participate in the research unless a person who is responsible for the health care of the subject has determined that the invitation to participate in the research and such participation itself will not interfere with the health care of the subject. . . .*

I have argued that in general, one should not invite patients to become research subjects without authorization of the physician responsible for their care.[8] However, this is the only recommendation made by the National Commission for a regulation that would require such consultation.

In the commentary under this recommendation, the National Commission further elaborates that when the potential subject's physician or other therapist is involved in the proposed research, independent clinical judgment should be obtained regarding the appropriateness of including that patient in the research. This is intended to reduce conflicts of interest between the objectives of health care and those of research, while still permitting clinicians, who may be especially knowledgeable regarding promising avenues of research, to apply their expertise in both enterprises. Though this recommendation is addressed to the same problem as Principle I.10 of the Declaration of Helsinki, it does not require that a third party obtain informed consent.

The National Commission recommended that in various situations there should be third parties in addition to the researcher and subject involved in the consent negotiations; in some circumstances they should also be involved in the continuing negotiations during the course of the research to see whether the subject wishes to withdraw from the protocol, among other reasons. In the National Commission's several reports, the third parties are variously called "consent auditors," "advocates," "neutral persons," and so on. Except for some types of research on those institutionalized as mentally infirm, the National Commission recommended that the need for such third parties be determined as a discretionary judgment of the IRB.

In my survey of the conditions under which third parties should be intruded into the relationship between researcher and subject, I used the following generic terms.[9] "Trusted advisor" is a term applied to those who act in an advisory capacity and who are or are not consulted according to the wishes of the prospective subjects or persons authorized to speak for them. "Overseer" is the term I use for agents whose employment is required by the IRB and who are empowered to prohibit the initial or continuing involvement of any particular subject.

When appropriate, there should be a suggestion that the prospective subject might wish to discuss the proposed research with another. When the proposed research entails a consequential amount of risk, discomfort, or inconvenience to the prospective subject, or when there are difficult choices between reasonable alternative therapies, consultation with a trusted advisor should be suggested, particularly if there are factors limiting the prospective subject's autonomy or capacity for comprehension.

Commonly, the trusted advisor is the prospective subject's personal physician when he or she has no involvement in the research. When the prospective subject has no personal physician or when the personal physician is involved in the conduct of the research, it might be appropriate to offer the services of another physician. In other cases, depending upon the nature of the problem, the prospective subject might wish to consult a trusted minister, lawyer, some other appropriate professional advisor, or even a friend who need not be a professional.

Suggesting consultation with a trusted advisor is quite a different matter from commanding the presence of an overseer. The requirement for an overseer should never be imposed frivolously. It is an invasion of privacy. The magnitude of the invasion can be reduced in some cases by allowing the prospective subject to select the overseer. Moreover, the imposition of such a requirement is tantamount to a declaration to the prospective subject that his or her judgment, ability to comprehend, ability or freedom to make choices, and so on, is to be questioned. However, in some cases, this will be necessary.

Resolution of the Conflict

I am inclined to agree with the National Commission that the dual role of physician–researcher should generally be permitted. There are conflicts, but these can usually be resolved. I shall discuss some approaches to their resolution shortly.

First, I should make it clear that I have not rejected out of hand Beecher's proposal of having at least two physicians involved in the conduct of research—one playing the role of physician–scientist and the other whose primary concern is the well-being of the patient. We must take seriously Fried's argument that one of the burdens imposed by participation in most, if not all, randomized clinical trials (RCTs) is that the subject is deprived of a relationship with a physician that is characteristic of medical practice—a physician whose only professional obligation is to the well-being of the patient, not complicated by competing obligations to generate high quality data. In Fried's words, the patient–subject is deprived of the "good of personal care."[10] Thus, I have argued that in programs like RCTs in which there is a prolonged exposure both to research and therapy—either validated (standard) therapy or nonvalidated therapy (e.g., investigational drugs)—one should take seriously the proposition that there ought to be a separation of the roles of physician and researcher.[11] In such cases though it might be quite appropriate to rely on the physician–researcher to provide day-to-day medical care, it might also be of value to offer to the patient–subject the opportunity to maintain a physician–patient relationship with a physician not involved in the RCT, but sufficiently familiar with it to facilitate the integration of its components and objectives with those of personal care.

I also agree with Beecher that informed consent is not merely not the only issue, it is not necessarily even the most important issue. There must be reasonable assurance that the patient–subject has ample opportunity to exercise his or her authority to withdraw without prejudice; moreover, there must be reasonable assurance that some competent professional will continue to observe the situation having a primary interest in the well-being of the patient–subject.

Except in cases involving special problems, which are characteristic of the RCT, it seems reasonable to rely on the physician–researcher to provide adequate protection of the rights and welfare of the patient–subject. In the event the researcher (physician or other type of professional) is not the personal physician, there should be a general presumption that no patient will be approached with an invitation to become a subject without the approval of the personal physician. Any exceptions to this general rule require justification—e.g., in some studies of the doctor–patient relationship, though it is essential to get the approval of the personal physician for approaching his or her patients in general, it may also be essential for the personal physician not to know which patients are being studied.[12] For another example, the intrusion on the doctor–patient relationship may be so minor that specific approval by the personal physician for involvement of each and every subject may serve no interest that justifies the expense and inconvenience—e.g., some types of studies of medical records and pathological specimens.

When confronted with a proposal to begin a project in which professionals will play the dual role of physician–researcher, judgments about whether it is necessary to introduce a third party into the relationship between the professional and the patient–subject should be made by the IRB. In general, these judgments should be made at the discretion of the IRB. We should refrain from developing regulations that would deprive the IRB of the flexibility it requires to make sound judgments that are appropriate to particular cases and to the institution it serves. In my view there are three factors that should be considered by the IRB in determining the necessity for special procedural protections. To the extent that any one of these three or any combination of two or more seems to present a problem, the IRB should consider it increasingly important to recommend special procedural protections such as trusted advisors or overseers.

1. The extent to which the prospective subjects have impaired capacities to consent must be considered. Are there serious limits to their autonomy, capacity to comprehend information, or are they legally incompetent?

2. The degree of risk presented by procedures performed in the interests of research should be taken into account. By definition, this means the degree of risk presented by maneuvers performed in the interests of developing generalizable knowledge. It does not mean the degree of risk presented by therapeutic, diagnostic, or prophylactic maneuvers—their status as "investigational" or "standard and accepted" notwithstanding. In considering whether the degree of risk is high enough to call for the consideration of special procedural protections, DHHS and FDA Regulations identify "minimal risk" as a threshold. The concept of "minimal risk" was developed by the National Commission as a threshold for determining the necessity for special procedural protections for children and those institutionalized as mentally infirm. As such, it is too low a threshold for autonomous adults. I have argued that a more suitable threshold for such persons would be a burden greater than what I have termed "mere inconvenience."[13] Research presenting mere inconvenience is characterized as presenting no greater risk of consequential injury to the subject than that inherent in his or her particular life situation. The risks that are relevant to these considerations are those of physical or psychological injury.

3. In protocols designed to introduce, test, evaluate, or compare therapeutic, diagnostic, or prophylactic maneuvers, there may be a need for special procedural protections to assure that: there is a clear and accurate statement of alternatives; the prospective subject will be afforded ample opportunity to make a valid choice between alternatives; and the prospective subject will be fully apprised of the consequences of choosing the dual role of patient–subject.

Summary

Inherent in the dual role of physician–researcher is a conflict of interest arising out of the competing objectives of research and medical practice. Most commentary and policy recommendations on this conflict of interest have focused on the problems that arise in negotiations for informed consent. However, these are not the only problems presented by this conflict; they are not necessarily even the most important. In order to deal with these problems, several commentators have suggested various procedural safeguards to protect the interests of the patient–subject—e.g., separating the roles of physician and researcher, introducing third parties into the relationship to assist in the initial or continuing negotiations for informed consent.

In my view, the necessity for special procedural protections for patient–subjects' interests should be a discretionary judgment of the IRB. In determining the need for special procedural protections for any research protocol, the IRB should consider three factors. To the extent that any one of these three or any combination of two or more seems to present a problem, the IRB should consider it increasingly important to recommend special procedural protections:

1. There are serious impairments of the prospective subjects' capacities to consent.
2. The risk of physical or psychological injury presented by procedures done in the interests of research is greater than "mere inconvenience."
3. The protocol is designed to introduce, test, evaluate, or compare therapeutic, diagnostic, or prophylactic maneuvers.

References

[1]R. J. Levine, *Ethics and Regulation of Clinical Research* Baltimore: (Urban & Schwarzenberg, 1981), p. 2.

[2]Ibid, pp. 5–6.

[3]Ibid, pp. 55–56.

[4]Ibid, p. 92.

[5]H. M. Spiro, "Constraint and consent: On being a patient and a subject," *New England Journal Medicine* 293, 1975, 1134–1135.

[6]H. K. Beecher, *Research and the Individual: Human Studies* Boston: (Little, Brown and Co., 1970), pp. 79–94.

[7]Ibid, pp. 289–290.

[8]Levine, pp. 103–106.

[9]Levine, p. 95ff.

[10]C. Fried, *Medical Experimentation: Personal Integrity and Social Policy* New York: (American Elsevier Co., 1974), p. 67.

[11]Levine, p. 130.

[12]Levine, p. 104.

[13]Levine, pp. 26–27.

Part 2

Personal Perspectives

Impact of Alzheimer's Disease and the Role of the Patient's Family

Hilda Pridgeon

Introduction

Alzheimer's disease. How prevalent is it? We are told there may be 1.5 to 2 million patients in this country. But let's take a broader look at how it affects families. Multiply 1.5 million or 2 million patients by 4, a conservative family size. That equals 6 to 8 million Americans who are closely involved as family members. Family members close enough to be deeply hurt by one of the most insidious and vicious diseases ever encountered by the human race. Humans have amazing recuperative powers. We can survive and even overcome major physical handicaps and go on to live triumphant lives. But what happens when the very tool needed to fight such disease and handicaps—the brain—is the first organ to fail? The Alzheimer's patient cannot fight it and the dismayed family must watch helplessly for years the all-too-slow decline of mental, then physical, capacity . . . some for as long as 12 to 14 years. We are all aware that the grieving process at the death of a loved one is a necessary passage, but what happens to families when grief becomes a daily companion over such extended periods of time? Indeed, the damage done by this monstrous disease may be more widespread than any of us realizes!

Case Histories

Although Alzheimer's disease happens most often to people over 65, let me give you some examples of families in Minnesota where we organized one of the first support groups for families of Alzheimer's patients.

In one family, the mother was about 35 when the first symptoms appeared for her. She has been diagnosed as an Alzheimer's victim and her husband has tried to continue life as the wage earner, be both mother and father to several young children, and cope with a wife and mother who is changing constantly on a downward course. The confusion for the children was a constant problem and family stress finally reached a point where the wife had to be placed in a nursing home.

The father of another family was 42 when a neurologist identified his strange forgetfulness, disorientation, and anger as Alzheimer's disease. His wife and two young sons went through a long series of bizarre behavioral episodes before finally seeking help from the Veteran's Administration Medical Center where he is now hospitalized.

A business executive began to miss appointments and show signs of confusion. His peers covered for him for awhile, but soon he was quietly moved out of his company to enforced retirement. His wife and teen-age children faced a bleak future trying to survive and pay for nursing home care. College plans may have to be delayed or abandoned.

A man in his late fifties retired early at reduced pension to care for his wife who is ill with Alzheimer's disease. He went through a staunch battle to care for her to the end at home, but when after five years of intensive 24-hour-a-day care, she finally needed a tube inserted to her stomach to avoid choking to death, he placed her in a nursing home where he visits her daily.

In another state, a 68 year-old woman cared for her husband until he became violent and she was forced to seek nursing home care for him. His Social Security goes for his care and since she has never worked outside the home, she was told, "You're on your own—go get a job." This, in an economy where *young* people are finding a scarcity of jobs! Eventually, their home was sold on a mortgage foreclosure and she worked in the home of a friend for her room and board.

One unusual and valiant lady is planning to care for her husband until he dies. He can no longer speak, but only laughs as a means of communication. She has approached the task in a matter-of-fact manner. She has special equipment to lift and move him, and a daily routine for his care. Usually incontinence, wandering, or the violence of a patient forces the exhausted caregiver to seek nursing home or hospital care.

These brief sketches should provide some idea of the range of families affected by what has been called the "silent epidemic." Alzheimer's disease is not just another "disease of the month" with which our legislators have become so familiar; it is most aptly described as "the disease of the century."

Stages of Family Coping

My own family has been dealing with the changing scene of my husband's Alzheimer's symptoms since diagnosis in early 1974 *and before.* Always there is that "before diagnosis" time. Most families will tell you that the gradual onset of memory loss and other symptoms began some time before a clinical diagnosis was sought or given. Most go through an initial stage of disbelief and non-acceptance of the hopelessness of the future for their patient. They may seek multiple diagnoses. Some may launch into a recommended health center program of concentrated vitamins, exercise, and special diets. Perhaps that type of program would not hurt any of us, but the tragedy is that the family is often told that Alzheimer's can be cured through such a program.

Families, and especially spouses, may doubt their own sanity at times, especially when friends or relatives say such things as, "But he looks so good and seems so well. Are you *sure* there's anything wrong?" Patients often retain the social skills the longest and carry off surface conversations for a long time.

I have been asked, "But aren't you glad for those years and months you didn't know what was wrong?" Glad for the times my husband accused me of lying to him? Glad when he lost his car repeatedly in large parking lots and we walked miles up and down rows of automobiles looking for it? Glad for his anger when he couldn't find things he had "put away," and then accused the family of stealing? Or glad to watch his attempts to hold onto a managerial job that had always been easy for him? Glad to watch a cheerful, gregarious, loving man turn into a confused, angry, withdrawn ghost of himself? No . . . not glad we didn't know. Certainly knowing what is ahead is appalling. Like a black pit at times. But not knowing can lead to such family distress that knowledge, when it comes, is actually a relief. "Ah, that explains why Dad did such strange things" or "That's why Mother forgot how to cook and couldn't find things in her own kitchen."

The second stage for families usually involves seeking others facing the same problems in order to compare coping techniques or frustrations. During this time, the knowledge of what the family faces descends upon them, and it can be terrifying. How long will it be before the patient cannot sign his name? What shall I do about property? Should I get a power of attorney or a guardianship? Many wives have never handled the family financial affairs or have never had to negotiate automobile repairs or a new roof on the house. These can loom as major stress factors in addition to the patient care. Families really need to get their financial affairs in order early in the course of the illness, particularly if the patient has been the major wage earner.

Other questions arising during this stage of family acceptance include, "Will we be able to care for the patient at home?" "Will she do damage to herself if left alone?" "Will violence be a factor?" "How much does Dad really know about what is ahead?" "Should the patient be told?" "How long should Mother be permitted to drive?" A major question asked by spouses and children alike is, "How long will Dad know who we are?" Those family bonds of love are so important to all of us. The spouse and children of a patient go through a real grief experience even in the early stages of the disease.

Another question that always arises is . . . "Is the disease genetic?" "What is the risk that siblings or children of the patient will be victims?" Some families have seen multiple cases of Alzheimer's disease and the remaining members live in fear as they approach their 40s and 50s. These families look to research with almost frantic impatience. Other families look with equal panic for some drug treatment that will help, and most are willing and eager to assist with research. Few really hold much hope for their current patient, but look to research to provide answers for the new generations.

Current Research

To all of us it seems incredible that only $10 million is being spent on research into a disease that kills so many. Or that research funds are less than 1/2000th the annual cost of caring for

Alzheimer's sufferers. If two million young people between the ages of 18 and 26 were dying with a disease, I feel sure this country would marshall a much greater force to find answers to the cause and a possible cure or treatment.

But let's not mislead ourselves . . . this nation is already paying a huge price in not attacking Alzheimer's disease with more research resources. It is a little like a commercial on television some time ago in which the auto mechanic points out the need for proper maintenance and repair and ends with a shrug, "Pay me now or pay me later," with emphasis on how much bigger the bill will be if the auto owner waits. We, as a nation, are paying nursing home care costs for Alzheimer's disease estimated in the range of $12.5 billion annually. This cost goes up each year as more patients are admitted and inflation takes its toll. In the next decade or two, when the over-65 population is projected to double or triple, the care costs for Alzheimer's disease could range between $25 and $35 billion a year.

To bring this monumental cost down to a more understandable level of one-patient/one-family range, the annual cost of nursing home care for an Alzheimer's patient is around $25,000/yr. Health insurance that the patient may have paid for years will not cover most of this cost. It is termed "custodial care" and every insurance company writes itself out of that type of benefit. Families who may have felt they had good health coverage find that the patient should have been more careful to get a disease that would be covered by their insurance. Even Medicare will do nothing to help families unless there is an acute nursing care function involved and then they will cover only a portion of that. What middle income family or spouse can long pay such annual costs and survive?

Financial Survival

Most families must seek state medical assistance with its stringent regulations—some requiring the selling of homes that, because of inflation, are now valued at too high a level to qualify under the law. Or they must sell an automobile worth too much to qualify, only to purchase an older model with continuous repair bills. Some of our state laws on medical assistance are not only cruel, but do not make financial sense. The spouse who must submit to welfare and medical assistance loses control over his or her own life simply to obtain care for the patient. In some states the only real answer to financial survival is to divorce the patient and force the state to help with medical care. But families do not seek this type of solution as a rule, and have a strong desire to care for the Alzheimer's victim.

When my husband was diagnosed, I was advised by the doctors and financial advisors to go back to his company where he had worked 25 years and seek reinstatement of disability benefits, pension, and insurance policies. He had resigned his management job that had become impossible for him, and thus had sacrificed all benefits. With the help of my company, Control Data Corporation, and their Employee Advisory Resource Department, we began the task. Even though his former company did not oppose the restoration of his benefits, it took two-and-one-half years of telephone calls and correspondence from an attorney to the insurance companies involved and the threat of a court battle, to obtain reinstatement.

Later, in our attempts to keep him at home as long as possible, we found a day care center that would accept him on a daily basis. Although the Sister Kenny Day Care Center in Minneapolis deals mainly with stroke and accident victims confined to wheelchairs, they accepted him on a trial basis. He was still physically strong, but the memory loss was acute. They discovered that he could pass lunch trays and help others, and when he pushed a wheelchair for someone unable to help themselves, if that person knew where he or she were going, they made a pretty good team. The family doctor and neurologist both wrote letters to Medicare and another health insurance company to indicate that this type of day care was beneficial to the patient and was indeed the only "prescription" that could help. The reply from Medicare was that the day care was "recreation for my husband and relief for his wife." Amazing that it could be classified as relief for me when I was out of the home working each day trying to keep the family together and a young teenage son in school. I requested a conference with the head of Medicare (Blue Cross–Blue Shield) in Minnesota and was refused.

Part of our battle to receive rightful disability benefits also involved insurance policies with waivers of premium. Such waivers are designed to maintain premium payments when a person becomes totally disabled. But we found that the fight to establish Alzheimer's disease as a true disability was long and expensive. Many insurance companies pretend not to hear the name Alzheimer's disease or to regard it as simple aging, and of course, not a disability. Even when doctor's statements emphasize that it is a progressive neurological disease, there is usually complete resistance. One family went to the New York State Insurance Commissioner to get a determination that Alzheimer's disease was indeed disabiling. However, both the New York case and my own are just the tip of the iceberg. The tragedy is that many elderly men and women, or even younger families, do not have the experience, preseverence, or resources to help them fight such battles, and so go down the drain financially and emotionally.

Emotional Stress

There is no reliable way to measure the emotional stress families sustain in the frustrating day-to-day care of a patient. I have spoken of the ongoing grief that engulfs them, but often one of the biggest stress factors is guilt. When human endurance reaches the breaking point, anger can be a natural result, and guilt a close follow-on emotion. When placement in a nursing home or hospital is necessary, the spouse wrestles with the "till death do us part" question and children who need to go on leading their own lives feel guilt for doing just that in the face of a parent's long-term illness.

I have spoken of the financial and emotional stress involved in dealing with patients with Alzheimer's disease. If a family or a spouse must deal with these problems and emotions alone, it can be devastating. I was involved in the formation of a family support group in Minneapolis/St. Paul in 1979 that is now the official ADRDA chapter in our area as well as regional headquarters. Later, we met in Washington with other such groups to organize the national association late in 1979. Since then, 87 chapters have been added in major cities around the country in 36 states. Others are in the formation stages. These chapters are gathering families together on a monthly or weekly basis to share resources and coping techniques, plan for education in their communities, and learn about research efforts.

The national office located in Chicago coordinates these efforts and handles national fund raising and publicity programs. A Medical and Scientific Advisory Board headed by Dr. Robert Katzman includes many of the leading researchers in the field and has launched a program of starter grants to encourage more research. We are pleased by the progress made thus far, but there is such a tremendous need in so many areas that we feel we have only begun. Our support of research efforts is wholehearted, but at the same time we face the problems and concerns of patients and families and the continuing long-term care needs.

Although this conference is directed to the question of informed consent, I was asked to speak on the impact of the disease on families. Our organization will be most interested in ways we can assist in tackling the problems researchers face in the informed consent question. When my husband was diagnosed and we were told of the prognosis, his response was, "We've had a good life and I'll fight it as long as I can." We were able to keep him at home for six years and I feel that part of the reason was his "fight." In that early stage he also stated that he wanted to help in whatever way he could with research and he actually did participate in a research project in Minneapolis. I feel I can speak for many, many families in the ADRDA when I say, we want to help with research. Help us find a way.

Advocacy for Persons with Senile Dementia

Nancy C. Paschall

In its classic sense, advocacy means to call to one's aid or to summon to one's assistance. Individuals with senile dementia of the Alzheimer's type (SDAT) have a special need for help in asserting their rights, a need that stems from three sources not dissimilar to those which also affect clients of mental health services (Kopolow, 1982). First of all, the nature of the illness itself makes it difficult for them to articulate their needs effectively. Secondly, the stigma attached to being a SDAT patient can lead to a tendency on the part of others to prejudge the capacity of patients and to underrate both their ability to function outside of an extremely controlled environment, and their ability to make decisions for themselves. Another stigma-related problem is the low priority given to patient concerns, simply because they are patients. Far too often, their wishes are denigrated, ignored, or treated as the ramblings of children who do not really know what is good for them.

An additional special problem of the SDAT patient is the difficulty of maneuvering through the incredibly complex support system that is created for the handicapped person, but for which few road maps have been developed. In addition to the need for advocacy to deal with the vulnerability issues and special problems of the individual, there is also a need for the change agent, monitoring force, or watchdog to bring about the creation of responsive services.

In short, advocacy on behalf of persons with SDAT is necessary. But, you may ask, don't we already evidence our concern for patients' rights through ethical codes, by the existence of institutional review boards, and in the regulatory machinery? These help, but they go only so far. All staff can see themselves as advocates for patients, but this is an attitude, not a function.

Family members can and are very effective advocates for their loved ones affected with SDAT. However, they cannot spend all of their time doing this; they must also maintain the rest of their lives.

Committees such as Institutional Review Boards (IRBs) and human rights committees are also important. The former review research proposals to determine the probable impact on human subjects, and are sometimes charged with intermittent review as the research progresses. In fact, however, the IRBs do not generally keep tabs on research once it has begun. The protection they afford, then, relates primarily to proposed activities, not to what may happen to subjects during the course of the research itself. Human rights committees are also important entities. Composed primarily of professional and citizen members from outside the caregiving agency, these committees already exist in many medical, mental health, and mental retardation settings. Examples include institutional review boards that oversee human subjects research, court-appointed boards designed to monitor implementation of court orders, and committees set up by institutions themselves to monitor client rights (Griffith and Henning, 1981). Again, however, this protection is one or more steps removed from the daily experience of patients.

Internal Rights Protection Advocates

Many types of advocates can assist SDAT patients. One of the most often utilized is the in-house rights protection advocate. Internal rights' protection advocates are on-site persons, usually hospital staff members, who educate patients and staff concerning rights issues, and who investigate complaints of rights' violations. They are available on the spot, are close to the situation of the in-patient, and know how to work the system administratively in a way that allows for quick, low-key education of patients and staff, and early attention to problems. Such advocates are now found on the staff of most mental health facilities and as patient representatives at many general hospitals as well. Their functions are many:

1. Helping patients to learn their rights through brochures, conversations, and signs around the ward.
2. Training staff concerning the rights of patients.

3. Setting up and implementing grievance procedures.
4. Serving on human subjects committees and reporting to institutional rights committees.
5. Being involved in the process when a patient must relinquish decision-making authority to another and reviewing decisions made by such surrogate decisionmakers.

Internal advocates, like other advocates, subscribe to the rule that they do what the patient requests. Their role is to help the patient obtain what he or she wants, to amplify the voice of people who otherwise might not have their opinion heard. Instead of doing what the advocate might think is in the best interest of the client, the advocate acts as the client directs. The bottom line is to assist client self determination (*Toward a National Plan for the Chronically Ill, 1980;* Scallet, 1977.) This can sometimes lead to friction within an institution, although the enlightened administrator will appreciate that adequate representation of patient views ultimately leads to better care. This attitude is exemplified by a recent statement from the Minnesota Assistant Commissioner of Mental Health who argued that advocacy should not be viewed as an extra financial burden, but rather as an integral part of the service system:

> *The advocate's position is perceived by the Mental Health Bureau as a direct care position because the advocate deals with individual patients/residents, their treatment and hospitalization programs, and other conditions and situations directly affecting individuals or groups of clients. The advocate's work is also related to the institution's quality of care. He/she is in a special position to hear of problems involving clients and their institutionalization, and to see that these problems are appropriately dealt with. Such activities not only protect individual clients but also help the facility maintain programs at the highest level. Aside from direct benefits to clients and programs, the hospital benefits through the early identification and correction of problems before they reach the stage of legal action, i.e., the risk management approach. ("Advocacy. . . .", 1981, p1.)*

Legal Advocates

Not all rights issues can be solved internally. Sometimes legal action must be taken, or at least threatened. Legal advocates, usually located organizationally outside the service system, represent individuals or classes of patients in litigation, or in other judicial settings short of actual litigation. By their position outside the service system, they are less open to real or potential conflicts of interest. Moreover, they can work in tandem with internal advocates to assure patients' access to representation both on a day-to-day basis within the facility, and intermittently through the courts when necessary.

Three nationwide advocacy programs may be available to SDAT patients. These are the Long-Term Care Ombudsmen Program, the Protection and Advocacy Program, and the Legal Services Corporation.

Long-Term Care Ombudsmen

SDAT patients who are senior citizens residing in nursing homes are eligible for assistance by advocates from the Ombudsmen Program of the Administration on Aging ("Long-Term Care Ombudsmen Program", 1981).

The 1978 Amendments to the Older Americans Act required that every State have an ombudsmen program. Specifically, Section 307(a) required that the State plan:

> *(12) provide assurances that the State agency will—*
>
> *(A) establish and operate, either directly or by contract or other arrangement with any public agency or other appropriate private non-profit organization which is not responsible for licensing or certifying long-term care services in the State or which is not an association (or an affiliate of such an association) of long-term care facilities (including any other residential facility for older individuals), a long-term care ombudsman program which will—*
>
> *(i) investigate and resolve complaints made by or on behalf of older individuals who are residents of long-term care facilities relating to administrative action which may adversely affect the health, safety, welfare, and rights of such residents;*
>
> *(ii) monitor the development and implementation of Federal, State, and local laws, regulations, and policies with respect to long-term care facilities in that State;*
>
> *(iii) provide information as appropriate to public agencies regarding the problems of older individuals residing in long-term care facilities;*
>
> *(iv) provide for training volunteers and promote the development of citizen organizations to participate in the ombudsman program; and*
>
> *(v) carry out such other activities as the Commissioner deems appropriate.*

As this book goes to press, the Act is up for re-authorization by the Congress.

Protection and Advocacy Agencies

Another advocacy resource for persons with senile dementia is the Protection and Advocacy (P&A) Agency that exists in every state receiving federal funds for services to developmentally disabled persons. By law, P&A agencies have jurisdiction to assist developmentally disabled persons in obtaining needed services, in resolving grievances, and, where necessary, in litigating. Although the agencies are not primarily legal in orientation, they all either have attorneys on staff, or can secure legal services for their clients. By definition, a developmentally disabled person must have a disability that began before the patient was 23 years old. However, advocates at P&A agencies can sometimes take on other clients, or can at least serve as good referral sources. The P&A agencies are not part of state government, but can be accessed by contacting the State Developmental Disabilities Program Office.

Legal Services Corporation

The Legal Service Corporation (LSC) is a federal entity that funds legal services for low-income persons through its grantees nationwide. LSC grantees originally concentrated on poverty law, but have recently branched out into other areas. As indicated by the title, these are legally oriented advocates, although their advocacy need not always result in litigation. The name and number of the local LSC office can be obtained through the national office of the Legal Services Corporation in Washington, D.C.

Self Advocacy

After having given an extensive argument on the need for advocates for SDAT patients, it may seem incongruous to recommend self-advocacy. However, clinicians (e.g., Kahn and Toben, 1981) are beginning to emphasize that doing too much for SDAT patients, infantilizing them, results in greater functional disability than is warranted by their organic impairment. On the other end of the spectrum, Verwoerdt (1981) warns against "therapeutic nihilism," another form of prejudging that is too easily adopted by persons working with senile dementia cases. He writes, "Whereas overinvolvement is an 'occupational hazard' in the treatment of children and younger adults, the other extreme, defensive withdrawal, is more likely to occur in clinical work with old patients." (pp. 198–199) In fact, however, this nihilism that leads to merely custodial care is unwarranted. Psychotherapy can be helpful in working towards the goal of re-establishing psychological equilibrium and maintaining contact with reality.

At both extremes, SDAT patients are harmed, either by being infantalized or ignored. In both cases, the unspoken message is that they are not full human beings. In fact, this is a form of prejudice, a prejudice that easily becomes part of the patient's own self concept, so that he or she begins to withdraw from all decisionmaking and activity, no matter how simple. This prejudice is reminiscent of that which plagued the mental health field prior to the emergence of the patients' rights movement in the 1960s and 70s. It is easy to conclude that this person must be totally "done for" and is incapable of independent choice, especially when that is the apparent personal belief of the patient.

Self-advocacy, especially on the part of patients in the early stages of SDAT, makes sense. Self-advocacy encompasses not only patients speaking their own minds, voicing their wishes, and pursuing their grievances. It also means that patients have a role to play in advising caregivers, researchers, and administrators on what it is like to be a patient, what it is like to be treated in a certain way, what the impact is on patients of various policies.

In mental health, patients have begun to take on a variety of roles beyond their "patienthood" (Paschall, 1981). This includes performing the advocacy functions at a Veterans Administration hospital (Manasse, 1981), designing evaluation tools at a psychosocial rehabilitation center (Prager and Tanaka, 1980), and serving on mental health advisory boards (Dyson, 1981). In other settings, current and former patients in the mental health system have served as trainees for staff members, by giving invited lectures, and making training videotapes. The most widespread form of client involvement is that of self-help/mutual-help groups. These range from organizations such as Recovery, whose members meet primarily for mutual support, to those such as the Network Against Psychiatric Assault, which is involved in radical political action. Although SDAT patients, especially those in the advanced stages of the disease, may not be able to function as autonomously as do the members of these groups, it should be remembered that the same attitude of "they can't do it" used to pervade the mental health area. Two conclusions emerge: (1) Some SDAT patients are able to function in other roles, and should be encouraged and assisted in doing so; (2) all patients should at least be given the opportunity and encouragement to act as self-advocates to the best of their abilities.

Rights Issues in Research

One means by which SDAT patients can hopefully help themselves and others is through participation in research. There are, however, a number of precautions that must be taken to assure that this participation is not harmful to the patient. Whenever possible, informed consent to the research should be obtained from the patient. The information should be given in such a manner that the patient can truly understand it, including using special means of communication to counteract memory loss, hearing or vision impairment, and the need for extra time to process the information. Further, precautions should be taken to avoid not only real concern, but also preceived threat—e.g., the fear that if one does not cooperate, medical care will be withdrawn.

It must not be believed that research is research is research. Patient involvement as an interview respondent is considerably less traumatic than that involving spinal taps or other intrusive procedures. Oversight should increase with the level of intrusiveness and possible negative consequences.

Finally, it must not be assumed that SDAT patients, even those with some substantial impairment, are unable to make their own decisions regarding participation. As several authors (e.g., American Health Care Association 1981; Gert and Culver, 1981; Katz, 1981) have written, a person's competence or incompetence to undertake one kind of endeavor (e.g., make legal decisions) does not necessarily equate with his or her competence to do other things, like agree to research. The draft report of the American Health Care Association's Ad Hoc Group on the

Problems of Questionably Competent Long Term Care Residents (Aug. 6, 1981) holds:

> *True total incompetency is usually found only in a comatose or severely ill individual. Most individuals in the population being discussed display intermittent and/or selective incompetence and are able to continue and to make and carry out some decisions. Because this ability should be respected and encouraged, the surrogate should make decisions only when the individual cannot or will not make them. Even then, the surrogate should endeavor to ascertain the individuals preference and/or assist him/her in making and carrying out decisions. (p. 5)*

Moreover, Katz (1981) has observed, ". . . competent and incompetent functioning cannot be neatly assessed. It is always a question of more or less, of one *and* the other, aggravated and attenuated by internal and external factors. Since external factors affect the balance between competence and incompetence, and do so to significantly differing degrees, the question must always be posed: incompetent for what external purposes?" (p. 104). He goes on later to note that, "What has been overlooked is that competence is not a fixed personality characteristic. The context in which it is evaluated has a significant impact on the conclusions reached." (p. 106). This means that the researcher or caregiver may be able to modify the SDAT person's environment in such a way as to assist him or her to make competent decisions. A number of suggestions present themselves for protection of SDAT patients participating in research:

1. Beware the "easy mark." If it seems that using this patient or class of patients is much simpler than is the case for other research subjects, that should serve as a red flag for the researcher. Are they being used only because they are passive? Has enough care been taken to explain the research to them?
2. Consider involving SDAT patients not only as research subjects, but also as paid consultants to the project in a variety of roles. This will, of course, pertain especially to persons in the early stages of the disease. They can preview the planned use of subjects to alert researchers as to how patients might perceive and respond to the procedures. They can assist in authoring the consent form. They can serve as communicators to potential subjects regarding the project, which may help to decrease perceived coercion.

3. Assist patients to make competent decisions by explaining the research to them in ways that are meaningful to them, at a pace that they can tolerate, with whatever personal or written assistance as can be had. This may include second visits or reminders.
4. In cases when a surrogate decisionmaker must be utilized, try to use someone as close as possible to the patient's situation. A relative may be a good choice if that person is close to the patient at the time of the decision. However, if the "closest" relative is hundreds of miles away and has not had contact with the patient for years, a more adequate surrogate decision might be made by staff, the advocate, a court, or some group close to the patient's living situation.

Conclusion: The Patient as Person

Many of the lessons and models discussed in this paper have been drawn from the experience with advocacy for the mentally ill and mentally retarded. Although the concept has, over the past decade, gradually gained acceptance in mental health and retardation settings, it is still a relatively new one concerning persons with SDAT. It can be expected that it will require fine-tuning to adapt to the specific needs of this group of people. However, the underlying concept is the same. The advocate takes as his or her organizing rubric the full-time concern for the client's human, legal, and clinical rights.

Working or living with persons with SDAT is draining. It is easy to deal with one's own frustrations by distancing the person. The following poem was found with the belongings of an elderly woman who died in the geriatric ward of a hospital near Dundee, Scotland. The poem refers to the resident of a geriatric ward, but applies easily to other persons—the mentally ill, the mentally retarded, and persons in the advanced stages of SDAT. It serves to remind us of the essential humanity of even the most seemingly deteriorated of persons.

What do you see nurses, what do you see?
Are you thinking when you are looking at me—
A crabby old woman, not very wise,
Uncertain of habit, with far-away eyes.
Who dribbles her food and makes no reply
When you say in a loud voice—"I do wish you'd try."
Who unresisting or not, lets you do as you will,

With bathing and feeding the long day to fill.
Is that what you are thinking—is that what you see?
Then open your eyes, nurse, you're not looking at me.
I'll tell you who I am as I sit here so still;
As I do at your bidding, as I eat at your will,
I'm a small child of ten with a father and mother,
Brothers and sisters, who love one another.
A young girl of sixteen with wings on her feet,
Dreaming that soon now a lover she'll meet;
A bride soon at twenty—my heart gives a leap,
Remembering the vows that I promised to keep;
At twenty-five now I have young of my own,
Who need me to build a secure, happy home;
A woman of thirty, my young now grow fast,
Bound to each other with ties that should last.
At forty, my young sons have grown and are gone
But my man's beside me to see I don't mourn.
At fifty, once more babies play round my knee.
Again we know children, my loved one and me.
Dark days are upon me, my husband is dead,
I look at the future, I shudder with dread.
For my young are all rearing young of their own,
And I think of the years and the love that I've known.
I'm an old woman now and nature is cruel—
'Tis her jest to make old age look like a fool.
The body it crumbles, grace and vigor depart,
There is now a stone where I once had a heart;
But inside this old carcass a young girl still dwells,
And now and again my battered heart swells.
I remember the joys, I remember the pain,
And I'm loving and living life over again.
I think of the years all too few—gone too fast,
And accept the stark fact that nothing can last.
So open your eyes, nurses, open and see
Not a crabby old woman, look closer, see me!*

Acknowledgments

The author thanks the National Institute of Mental Health for support of the work reported here. The opinions expressed in this article are those of the author alone and do not necessarily represent those of the Institute.

*Courtesy of the Peninsula Hospital Center, Far Rockaway, New York, and Greater New York Hospital Association.

Notes and References

"Advocacy reaffirmed in Minnesota," *Au Contraire,* 2(4), 1981, 1.

American Health Care Association. Draft Report of the Ad Hoc Group on the Problems of Questionably Competent Long Term Care Residents. Washington, DC, August 6, 1981.

K. Dyson, "New roles for clients: the client as board member." Presentation to International Association of Psycho-Social Rehabilitation Services, Washington, DC, 1981.

B. Gert & C. Culver, "Philosophical Overview." Presentation to NIMH Workshop on Empirical Research on Informed Consent with Subjects of Uncertain Competence, Rockville, MD, 1981.

Robert G. Griffith and Dana B. Henning, "What is a human rights committee?" *Mental Retardation,* 1981, 61–63.

R. L. Kahn & S. S. Tobin, "Community treatment for aged persons with altered brain function," in N. E. Miller and G. D. Cohen, (Eds.) *Clinical Aspects of Alzheimer's Disease & Senile Dementia* (New York: Raven Press, 1981).

J. Katz, "Disclosure and consent in psychiatric practice: Mission impossible?" in Charles K. Hofling (Ed.), *Law and Ethics in the Practice of Psychiatry* (New York: Brunner/Mazel, Inc., 1981).

L. E. Kopolow, "Patients' rights and the therapeutic relationship," in *Rights of the Mentally Disabled,* Washington, D.C.: American Psychiatric Association, 1982.

G. O. Manasse, "Patient participation in facility management at a VA psychiatric hospital," *Hospital and Community Psychiatry,* 1981, 32, 871–872.

N. C. Paschall, "New roles for clients: implications for policy," Presentation to International Association of Psycho-Social Rehabilitation Services, Washington, D.C., 1981.

E. Prager and H. Tanaka, "Self-assessment: the client's perspective," *Social Work,* 25, 1980, 32–34.

L. Scallet, "The realities of mental health advocacy: *State ex rel Memmel v. Mundy,*" in L. E. Kopolow and H. Bloom (Eds.) *Mental Health Advocacy: An Emerging Force in Consumers' Rights* (Rockville, MD: National Institute of Mental Health, 1977).

"The Long-term Care Ombudsman Program: Development from 1975–1980." Washington, DC: Administration on Aging, 1981.

Toward A National Plan for the Chronically Mentally Ill. Washington, DC: Department of Health & Human Services, 1980.

A. Verwoerdt, "Individual psychotherapy in senile dementia," in N. E. Miller and G. D. Cohen, (Eds.) *Clinical Aspects of Alzheimer's Disease & Senile Dementia* (New York: Raven Press, 1981).

Ethical Issues in the Care of the Patient Involved in Alzheimer's Disease Research

Edward W. Campion

Introduction

To date, senile dementia of the Alzheimer's type (SDAT) has been found only in human beings, so it is axiomatic that Alzheimer's research must involve human subjects. These human subjects, of course, are patients with the disease under the care of physicians. If and when the SDAT patient becomes a research subject, complicated problems result concerning the rights of the patient and responsibilities of both physician and researcher. The first professional obligation of the physician is to the welfare of the patient and, particularly in Alzheimer's, to the patient in the context of the family.[1] There are, to be sure, other professional values to be promoted, including the development of medical knowledge, but the foremost obligation is to provide the best possible care of the patient, including the Alzheimer's patient who may be a research subject.[2] The interrelationships are complex between the duties of physican and researcher, and between the rights of patient and research subject.

Research and Patient Care

Exemplary care of the patient in context of SDAT research is more than a virtuous ideal. It is a basic necessity for several concrete reasons. First, the validity and reliability of the research may actually depend upon the patient receiving optimal care.

Patients must be properly diagnosed. Frequently overlooked treatable conditions that can mimic Alzheimer's or can greatly worsen the patient with only mild Alzheimer's must be recognized. These conditions include hypothyroidism, electrolyte imbalance, pernicious anemia, normal pressure hydrocephalus, drug intoxication, and depression, to name but a few. Second, medical complications frequently arise during the relentless course of SDAT. All too often these patients are prone to infections, aspiration pneumonitis, falls, malnutrition, and a host of diseases.[3–5] Given substandard care, the SDAT patient's course may become needlessly complicated and needlessly painful. Conversely, the SDAT research should be able to focus upon SDAT manifestations with the minimum possible number of complicating factors. No one wants to base research conclusions on poor medical care.

Third, the very future of SDAT research may depend more upon the clinical care of the research subjects than upon the quality of the research itself. The former is much more visible and easier to judge than the latter. In this area it has to be recognized that public opinion counts tremendously. There is, in mind of the American public today, more than a little ambivalence, suspicion, even distrust of the medical establishment—and even more so of medical research establishment. The worst case example may be instructive. Suppose that scientifically excellent research is undertaken that involves SDAT patients in a grossly substandard nursing home. Since the patients are incompetent to give informed consent, it has been waived.[6] The researchers keep their focus on their scientific pursuits alone. Patient care is left entirely to the nursing home staff with no questions asked. In such a situation the researchers, naively, are endorsing and supporting the substandard care. Their presence and their professional prestige together with that of their parent institution carry with it tacit approval and endorsement. This situation also carries all ingredients for public scandal. That may, in fact, be desirable if it exposes elder abuse, neglect, money gouging, and violation of the standards of decent human care. However, when the exposé comes (as it should), it will include the university and perhaps even the NIH funding source. In this scenario, the most assured result will be that the SDAT research enterprise in general will suffer, as will the individual researchers.

Patient Care and Ethics

Beyond the pragmatic and utilitarian ramifications, professional and ethical standards mandate that an acceptable level of human care care be firmly reaffirmed as a *sine qua non* for human research.[7,8] Patients with advancing SDAT are vulnerable to the point of helplessness. They cannot serve effectively as their own advocates. Families generally, but not always, try to act to protect the patient's best interests. In any case, families in this context are subject to a wide variety of coercive influences from the pressures of medical–scientific authority to the desire to cooperate uncritically for fear that anything less might jeopardize their family member's care.[9,10] This places the researcher in the uncomfortable position of having to judge the quality of care given by other persons and by other institutions.[11] Frequently the subjects are nursing home residents. Judging the quality of care is not a simple matter, but it is possible. Fine care of SDAT patients may avoid quantification or definition by a handy list of guidelines, but any clinician in the field can recognize it easily. Some kind of judgment of patient care quality must be made by the researcher both for ethical and practical reasons. To propose otherwise would be to designate a large group of helpless human beings as a class of research material without even the guarantees of basic care accorded to lower animals.[9] The worthy goal of the future conquest of SDAT cannot be used to justify research endeavors that in any way involve substandard care of human beings.

In a sense, research in the field of SDAT is high-risk research. Despite recently increased education efforts there remains great public ignorance concerning Alzheimer's disease. Research serves to bring the disease to light, including the awful tragedy of far-advanced Alzheimer's. The researcher, then, becomes the bearer of the bad news. Such messengers of bad news have been in a precarious position since at least the time of Sophocles. Hence, first-rate patient care again becomes essential for the very preservation of the SDAT research. The disease itself is bad enough without the complications of neglectful or even abusive care which has been condoned on some forgotten wards.[12] Moreover, during the course of any clinical research on SDAT, the patients sooner or later are going to worsen from the relentless progression of the disease. It requires effective medical and nursing care to detect any possible complications of the research, to distinguish them from disease progression, and to help the family keep the two separate.[13]

A further reason that SDAT research patients deserve excellent care is simply because *that* should in itself become one of the major issues in the research agenda.[14] How can the very best care be provided these patients? This question is of major concern to primary care medicine and to the growing field of geriatrics.[15,16] The focus of research has to extend to the practical care of the many patients suffering from what may remain an incurable disease for some years to come.[17] One means to this end is the stategy of developing centers that demonstrate and innovate upon optimal long-term care of SDAT patients. The teaching nursing home must also become the research nursing home.[18]

It must be stated frankly that, at present, getting first-rate care for SDAT patients is not at all easy. Long-term care is a side of medical care that is undertaught, undervalued, undersupported . . . and also underresearched.[19,20] The Alzheimer's patient may be forgotten and abandoned or may be inappropriately oversedated by medications. Ageism may be a major factor.[21] Patient and family may find care-givers, from physician to nurse's aide, manifesting attitudes of "Who cares?" "What can be done anyway?" . . . or worse. Moreover, we are entering an era of increasingly stringent fiscal restraints in medical care.[22] These restraints threaten to affect the elderly, poor, impaired patient the most severely. One cannot help but wonder if a message is being sent that resources should be shifted away from such patients, that the economically unproductive patient is to be kept out of the mainstream of medical care and research.

Patient Care Factors

Excellence in care for the SDAT patient is made difficult by the frequent complexity of multiple associated problems. Diagnosis becomes difficult and management is challenging in these patients. Communication may be impaired. Disease symptoms may be elusive.[23] Physicians and other health providers receive very little teaching and training in such areas. With the current emphasis on acute care, dramatic interventions, and technological agility, physicians caring for SDAT patients may become frustrated and confused concerning treatment goals. Some respond to this by aggressive, blind overtreatment without regard for goals of patient and family.[24] Some respond by refusing to do anything. Even conscientious, dedicated health professionals simply become burned out and exhausted in their work. It is a difficult and discouraging disease.

The best care for SDAT is that which involves all aspects of the patients' life, environment, and family. A team approach works best when goals are clear and humane, and communication is open and unhindered.[25] The care may be centered at home, in a formal home-care program, in foster care, in day care, or in institutional long-term care. Wherever it is, the care of the patient, even the medical care of the patient, should not be mistakenly equated with what transpires in the few minutes per month that is spent with the doctor. It encompasses an entire network of people and a strategy that must involve the patient's total environment.

A recurring major issue in the care of the SDAT research subject is simply "who cares for the patient?". . . or more specifically, should the patient's doctor be someone involved in the research or someone entirely apart? Properly conducted, major clinical research efforts have generally led to improved care as, for example, in hypertension and in cancer. The researcher has considerable interest in seeing that the patient is optimally cared for. The clinical investigator may also be more experienced, skilled, and dedicated to the care of the SDAT patient. However, the researcher–subject relationship is clearly different than the doctor–patient relationship.[26] When all rests upon a single individual there are definite dangers that a monopoly in decision making can lead to conflicts of interest and to a certain self-protective sense of infallibility.[27] On balance, both patients and research will be best served if there is some separation maintained between care-giver and researcher. This is particularly true of research carrying anything more than trivial risk. On the other hand, research that is virtually devoid of risk or discomfort may be considered appropriate to be conducted by those caring for the patient, with institutional and procedural safeguards.[28] To do otherwise will stifle the professional development of a group we vitally need—those professionals dedicated to the health care of the impaired elderly, including the impaired elderly Alzheimer's patient. In actual fact most clinical research involves very little risk. A 1976 survey reported that in nontherapeutic research on some 93,000 subjects, less than 1% of subjects suffered any injury at all.[29] Of those suffering some injury nearly 90% were trivial injuries such as local reactions, minor burns, mild allergic reactions. There were no fatalities and less than 1/10 of 1% sustained any injury resulting in disability. This is not, of course, to deny the necessity of safeguards and precautions. There remain research "risks" that are much harder to tabulate: hassle, discomfort, time, feeling like a guinea pig, and just plain worry.

Qualified clinical researchers are a precious resource, perhaps even an endangered species. Unwieldy and unreasonable requirements may discourage them from even entering the field of Alzheimer's research. Complex procedural formalities of consent become major hindrances. The fight against Alzheimer's will be the major victim. What truly matters is that the researcher be principled, sensitive, honest, and intellectually sensible. Procedurally perfect consent for a study that is scientifically nonsensical is, in itself, unethical. The concerned, conscientious researcher remains one of the best guarantors of ethically sound human research. Such persons will also improve the care of SDAT patients whether through direct responsibility or as part of a demonstration effort. In the coming era of increasing biomedical research by private industry, it becomes all the more essential to re-establish the primacy of the ethical and humane care of SDAT research patients.[30] It must be viewed as a first principle of human research, not simply as a consequence of federal funding guidelines.

Summary

Ultimately, the future care of SDAT patients and the hope for improving that care relies very largely upon research.[31] The costs of the disease are so staggering—in hard dollars and in human despair—that we cannot afford to let Alzheimer's research become stalled or, worse, to die the death of a thousand qualifications. No disease affects us more profoundly nor threatens us more tangibly than Alzheimer's. Because it is so common and so highly age-related, the care of the Alzheimer's patient is closely linked to the health care of the elderly in general, and of the seriously impaired elderly in particular. The simply stated imperative of the Golden Rule still stands as a first principle in both human research and patient care. It is particularly appropriate to our thinking about SDAT since in the years ahead those afflicted with the disease may potentially include any of us or our families. That chilling reality should serve as a stimulus to facilitate Alzheimer's research and as an impetus to develop superlative care for those patients.

Notes and References

[1]C. K. Cassel and A. L. Jameton, "Dementia in the Elderly: An Analysis of Medical Responsibility," *Annals of Internal Medicine* 94, 1981, 802–807.

[2]R. Goldman, "Ethical Confrontations in the Incapacitated Aged," *Journal of American Geriatrics Society* 29, 1981, 241–245.

[3]R. A. Garibaldi, S. Brodine, S. Matsumiya, "Infections Among Patients in Nursing Homes: Policies, Prevalence, and Problems," *New England Journal of Medicine* 305, 1981, 731–735.

[4]J. G. Ouslander, "Drug Therapy in the Elderly," *Annals of Internal Medicine* 95, 1981, 711–722.

[5]K. Steel, P. M. Gertman, C. Crescenzi, and J. Anderson, "Iatrogenic Illness on a General Medical Service at a University Hospital," *New England Journal of Medicine* 304, 1981, 637–642.

[6]A. Meisel and L. H. Roth, "What We Do and Do Not Know About Informed Consent," *Journal of the American Medical Association* 246, 1981, 2473–2477.

[7]C. B. Chapman, "On the Definitation and Teaching of the Medical Ethic," *New England Journal of Medicine* 301, 1979, 630–634.

[8]D. Callahan, "Shattuck Lecture: Contemporary Biomedical Ethics," *New England Journal of Medicine* 302, 1980, 1228–1233.

[9]H. K. Beecher, *Research and the Individual* (Boston: Little Brown, 1970).

[10]G. J. Annas, "Informed Consent," *Annual Review of Medicine* 1979, 29, 9–14.

[11]H. K. Beecher, "Ethics and Clinical Research," *New England Journal of Medicine* 274, 1966, 1354–1360.

[12]N. K. Brown and D. J. Thompson, "Nontreatment of Fever in Extended Care Facilities," *New England Journal of Medicine*, 1979.

[13]P. P. Vitaliano, A. Peck, D. A. Johnson, P. N. Prinz, and C. Eisdorfer, "Dementia and Other Competing Risks for Mortality in the Institutionalized Aged," *Journal of American Geriatrics Society* 29, 1981, 513–519.

[14]R. N. Butler, "Geriatrics and Internal Medicine," 91, 1979, 903–908.

[15]P. E. Dans, and M. R. Kerr, "Gerontology and Geriatrics in Medical Education", *New England Journal of Medicine* 300, 1979, 228–232.

[16]C. Eisdorfer, "Care for the Aged: The Barriers of Tradition," *Annals of Internal Medicine* 94, 1981, 256–260.

[17]P. B. Beeson, "Training Doctors to Care for Old People," *Annals of Internal Medicine* 90, 1979, 262–263.

[18]R. N. Butler, "The Teaching Nursing Home," *Journal of the American Medical Association* 245, 1981, 1435–1437.

[19]A. Leaf, "The Spector of Decrepitude," *New England Journal of Medicine* 299, 1978, 1248–1249.

[20]R. Kane and R. Kane, "Care of the Aged: Old Problems in Need of New Solutions," *Science* 200, 1978, 913–919.

[21]R. N. Butler, *Why Survive?* (New York: Harper and Row, 1975).

[22]L. B. Russell, "An Aging Population and the Use of Medical Care," *Medical Care* 19, 1981, 933–942.

[23]L. S. Libow, "Pseudo–senility: Acute and Reversible Organic Brain Syndromes," *Journal of American Geriatrics Society* 21, 1973, 112–120.

[24]E. W. Campion, A. G. Mulley, R. L. Goldstein, G. O. Barnett, and G. E. Thibault, "Medical Intensive Care for the Elderly," *Journal of the American Medical Association* 246, 1981, 2052–2056.

[25]L. S. Libow, "A Geriatric Medical Residency Program," *Annals of Internal Medicine* 85, 1976, 641–647.

[26]H. M. Spiro, "Constraint and Consent—On Being a Patient and a Subject," *New England Journal of Medicine* 293, 1975, 1134–1135.

[27]H. Brody, *Ethical Decisions in Medicine* (Boston: Little Brown, 1981).

[28]W. J. Curran, "New Ethical–Review Policy for Clinical Medical Research," *New England Journal of Medicine* 304, 1981, 952–954.

[29]P. V. Cardon, F. W. Dommel, and R. R. Trumble, "Injuries to Research Subjects: A Survey of Investigators," *New England Journal of Medicine* 295, 1976, 650–654.

[30]J. F. Childress, *Priorities in Biomedical Ethics* (Philadelphia: Westminster, 1981).

[31]J. W. Rowe, "Clinical Research on Aging: Strategies and Directions," *New England Journal of Medicine* 297, 1977, 1332–1336.

Part 3

Historical, Legal, and Ethical Background

Research Objectives and the Social Structuring of the Research Enterprise

An Historical and Ethical Perspective

Harry Yeide, Jr.

Introduction

This essay is written in the belief that current legitimations of and regulations for medical research involving human subjects are the outgrowth of a fairly long, if often forgotten, history. We will attempt to bring certain chapters of this history to light, partly in the hope of providing a useful orientation, and partly to suggest that our ethical tradition may be in need of some new insights and directions at this point in medical history.

Justifications for Social Permission to Experiment on Humans

Before presenting any of the historical monuments, I think it will be useful to remind ourselves of some of the possible legitimations for medical experimentation on human beings. The general acceptance of such experimentation in modern societies ought not blind us to the need for these legitimations. So far, modern societies have professed to believe that subjects must consent to participation in such experimentation because it typically entails some degree of risk and a temporary loss of power to the experimentors. I have encountered at least seven arguments that claim to legitimize medical experimentation on humans and that occur in various combinations with one another. It is my sense that the first three on my list are most common in current literature, but I have done no rigorous content analysis to verify that impression.

Social Utility

One often hears justifications of human experimentation based upon social utility. New therapies are an important social good both for afflicted individuals and for the community as a whole. Human experimentation is seen as an indispensible means to the development of such new therapies and is, therefore, justified. Since experimentation can also discover treatments that are inferior or even harmful, the use of human subjects in discovering "bad" therapies can also be portrayed as socially useful in reducing harm to the individual and the society. In some European literature produced in countries that are major pharmaceutical exporters, I have seen the additional observation made that human experimentation contributes in a substantial way to the economic prosperity of the society. Though this may strike some of us as a less lofty good than health, it is hardly out of place in a justification of human experimentation built on the basis of social utility.

Justice

Another kind of legitimation is built on a notion of justice. It is pointed out that all living persons are the beneficiaries of experiments done on those who lived before us. Our "debt" to them can be paid by offering ourselves as subjects in experiments that will benefit future generations. Presumably any immediate benefits from participation are seen as an extra bonus, for the main weight of this argument falls on the satisfaction of justice over several generations.

Freedom of Science and the Scientist

A rather different kind of justification can be built on the freedom of science and the scientist. Sometimes freedom of research is portrayed as the penultimate value, truth being the ultimate value to which it relates. But occasionally freedom of enquiry appears itself to be the ultimate value. The cause of truth and the activity of science are thought to be threatened by any limits whatsoever on investigation. Reference is often made to various episodes of totalitarian interference with scientific activity and the tragic consequences for science and truth. In a strong form, this rationale can set aside such things as consent requirements, protection of vulnerable populations, and so on. Generally, however, this view is conjoined at least with the requirement that undue injury not be done to human subjects. This kind of justification is probably more common among behavioral scientists than among medical experimentors.

Love and Justice

Another sort of legitimation is built on notions of love or justice, or a combination of the two. Stated somewhat formally, it holds that the competent (the healthy) have an obligation to help (heal) the less competent (the sick). Medical experimentation on humans is seen as an essential means to this end, and is justifiable even at the cost of some risk and harm. This may sound like the social utility argument, but is significantly different. If, for example, a particular level of medical care and the experimentation necessary thereto should no longer be seen as helping maximize social utility, the obligation to experiment would diminish or disappear for the social utilitarian. The point of view presently being described would continue to recognize the obligation to support care and experimentation even if this resulted in less than the maximum social utility. (This obligation would, of course, have to be weighed in relation to others.)

The Relation Between Science and Technology

Still another legitimation is derived from a set of impressions regarding the relation between science and technology. It is held that science and technology are in an asymmetrical relation to one another, that technology or applied science derives from pure science. If one treats medical care as an applied science, this suggests that it is in a fundamental way dependent upon the kind of pure science represented by at least some medical research. The inference is then available that if you want to enhance the sophistication of medical care (as contrasted with, for example, its general availability), then you must nourish its scientific base.

This line of argument is, in my opinion, more vulnerable to attack on the basis of certain facts and other interpretations of facts than is the case with the four foregoing justifications. It can be questioned whether very much medical experimentation is "pure science" since so much of it is conducted to answer particular therapeutic questions. And at least since the major writings of Jacques Ellul, there have been those who see science as the consequence rather than the cause of technology. Even if the notion that science gives rise to technology is generally true, medical science may differ in significant ways that make it an exception to the rule. But vulnerable or not, it is an approach that I encounter from time to time.

Sacrifice

Another justification for medical experimentation on humans can be built on the notion of sacrifice. It appears universally to be the case that societies discern some situations in which human sacrifice is legitimate, even obligatory. Though the phrase "human sacrifice" may most quickly bring to mind primitive priests slaying their victims on stone altars, the more common sacrificial situations for societies of all kinds are probably war and famine. It is interesting that in the United States, those who have been exempted from one situation of human sacrifice (i.e., war) on the basis of their conscientious objections to violence have often been obliged to participate in another situation of human sacrifice—medical experimentation. Although I have never seen printed documentation, I have been told by several in a position to know that some of these "volunteers" did indeed sacrifice their lives. This rationale for medical experimentation has not been greatly discussed in literature with which I am familiar, but it is a central theme in two of the essays that I regard as among the most profound treatments of this problem in the 20th century—those of Viktor von Weizsaecker and Hans Jonas.[1] Both find reasons for holding that an experimental subject must freely will participation in this form of sacrifice; the community cannot demand it in any typical situation. However, it is held to be moral for the community to request such sacrifice.

Anthropological Justification

The last line of legitimation to which I would draw your attention is what might be called an anthropological justification. It grows out of a particular portrayal of what it means to be human, and argues that the human is by nature and destiny an experimental animal. We are reminded that humans are instinct-poor, that we are required to experiment with our world and ourselves if we are to exist at all. Our highest human achievements are the consequence of noble experiments. What we call science is the modern arena in which this primeval human reality chiefly expresses itself. What we call medical experimentation calls into play another fundamental human possibility—care. The medical experiment as seen from this perspective appears as an occasion for the supreme expression of humanity on the part of the experimentors and the subjects. Only weighty considerations would seem to justify nonparticipation. There are a number of philosophical languages in which this view can be set forth. In this country, the thought of John Dewey represents such a vehicle.[2] Although not the only legitimation that might do so, this view can be especially supportive of the notion that experimental subjects are co-investigators with those who design and supervise the experiment.

Emerging Moral Boundaries

With our inventory of possible legitimations in mind as points of orientation, let us examine a sample of past reflections on the ethics of experimentation, confining ourselves to relatively modern times. Generally, these statements have assumed one or more of the foregoing legitimations and added one or more other moral criteria thought to be necessary if human experimentation is to be kept within acceptable moral boundaries. One of the most frequent additions has been the demand for free, informed consent, which reminds us of a central concern in this conference. However, consent issues have not always enjoyed the central status that they have in recent American discussion. A possible dividend in attending to the longer historical tradition might be some insight into the problem of balancing consent issues with other equally important concerns.

It should be made clear at the outset that your reporter is not an historian of medicine, and is not in a position to inventory all or even most of the past wrestling with the question of human experimentation. Most of what he has to report comes from the German tradition. Given the worldwide influence of German scientific medicine in the late 19th and early 20th centuries, it is very possible that this sector of world medicine also produced the most advanced analyses of the topic. In any case, Albert Moll, who will be our main witness to the state of the discussion in 1902, already knows of an effort at public regulation by the Prussian Minister of Culture (which would have had force in a large part of pre-World War I Germany), and of a proposal from 1838 by a Leipzig investigator that the problem of obtaining informed consent be solved by establishing an organization of physicians who would make themselves available for drug trials, though only in expectation of a "truly very great utility."[3]

But before reviewing some of Moll's own conclusions, let us recall the oft-cited contribution of Claude Bernard. Bernard has been credited with enormous influence regarding the ethics of human experimentation.

> *The first express formulation of the scientific rationale for human experimentation was that of Claude Bernard's famous Introduction to the Study of Experimental Medicine first published in 1865. Although Bernard's scientific approach to medicine was widely adopted and further refined, his ethical principles tended to be accepted as presumptions so basic as to require little, if any, discussion. This ethical equinimity of the medical profession was shattered by the reports from Nazi Germany of atrocities committed in the name of clinical science.*[4]

The comments dealing with the ethics of human experimentation are found in that part of the book that deals with vivisection. Since there was a raging debate on that topic around the turn of the century, I find myself doubting that there was universal acceptance of the moral principles by which Bernard justified that kind of experimentation. One might even wonder how well-known his thoughts were in the English-speaking world, since the *Introduction . . .* was not translated until 1927.

Be that as it may, his discussion is primarily a justification of vivisection using both humans and animals. He reviews the antiquity of the practice of experimenting on humans—especially on criminals and the condemned. He notes how resistence to such work has blocked the advance of scientific medicine. Such experiments on humans are justified as extensions of normal medical practice. Indeed experimental vivisection is portrayed as an obligation "whenever it can save his life, cure him or gain him some personal benefit." He reports himself as opposing experiments on the condemned, but then approvingly reports a study involving the introduction of experimental larvae, without the subject's consent, into a woman who will be examined after her execution. Some would view that as too limited a concern for the condemned, not to mention the canons of consent. He does indicate allegiance to the principle that we should not harm the neighbor. He expresses the opinion that too many dangerous experiments are performed on humans, which he conjoins with his confidence that better animal experimentation will render them unnecessary. He also suggests that in weighing the permissability of various experiments, the only court of appeal should be one's own conscience or other scientists. My own sense is that his most basic legitimation is the "anthropological justification," revealed in his comment that "in everyday life, men do nothing but experiment with one another."[5] In short, Bernard offers strenuous justification for experimentation on humans, but very little help deciding how such experiments might be kept within moral boundaries beyond the admonition to avoid experiments that can only harm the subject.

Let us turn now to Moll's work of 1902. Over 100 of its 650 pages is devoted to "medical science and research." Though he is concerned with a wide spectrum of medical ethical issues, it is in part as a response to unethical experiments that he undertakes this massive effort.

> *Precisely the circumstance that physicians raise so little protest . . . when humans are treated as guinea pigs, while any violation of the duties of the status group are seen as an offense or injury to the profession: all this proves how necessary it is to investigate the true ethical duties of the physician.*[6]

Moll appears to have worked through a substantial amount of the ethical literature of his day, though his disdain for citing sources makes his pilgrimage impossible to follow. Suffice it to say that he denies being a utilitarian, a believer in the ethics of universal evolution, and quite a number of other things. His point of departure is an inventory of moral feelings, and he thinks it a matter of indifference whether they be placed in us by God, socialization, or whatever. He also knows that it requires art appropriately to connect these feelings to given situations and decisions.

He also takes seriously the way that participation in a profession might condition moral feelings. This is not for him, as it would be for some philosophers, a problem to be overcome, but a moral reality of the first magnitude. Thus a physician has one group of moral feelings that are universally human. But:

> *The second group is derived from the profession of the physician. I regard as belonging thereto the right to heal the sick and to maintain the life of the cripple, and this right is an unprovable presupposition. (p. 16)*

Somewhat earlier, Moll also declares a right of the ill to receive care (p. 8).

Although life in the profession ideally strengthens these moral feelings, it is also possible to lose the way and to allow other considerations to displace what is crucial to the profession.

> *Thus we find that the violation of basic ethical principles by physicians is often the consequence of their special activity. Whoever is totally or mainly oriented toward scientific research forfeits relatively easily the physicians' way of thinking. (p. 31)*

Thus the issue of human experimentation is a problem for physicians as well as subjects in Moll's view. And it is clearly more than a problem of custom, for he takes great care to distinguish his moral considerations from such things as etiquette, policy, tact, custom, and the like. Widespread practice does not render an activity moral.

When he returns to the topic of experimentation some 400 pages later, he repeats his sense of distinction between medicine and natural science.

> *Medicine as such really pursues a purely practical goal . . . battling sicknesses. If one counts medicine as a natural science, that is only correct to the extent that medicine studies the human as a natural body in order to achieve its aim. Learning about the characteristics of the human and the controlling natural laws is not in itself a goal. (p. 474)*

Although Moll affirms the marriage of science and medicine, he feels that extravagant claims for science have long been a strain in the marriage. (He introduces some spokesmen from the 18th century who already thought that the work of science was largely complete.)

Moll discusses and approves animal experimentation and autopsy before moving to experimentation on living humans. He offers his comments as a response to some 600 experiments he has found described in the professional literature, all of which lacked any likely therapeutic benefit for the subjects. He is convinced that this is just the tip of the iceberg. Unfortunately, he again fails to provide citations that would allow us to reconstruct his survey; he may have felt that professional bonds demanded such discretion. But there are sufficient clues to make it clear that he is discussing experiments in many nations. It is probably also unfortunate for the modern reader that his morally significant judgments are dispersed amidst the discussion of the cases. So you must read the whole, or rely on my summary. Applying consistently his conviction that his profession is committed to healing rather than knowledge, that there is a right to heal and a right to be healed, he derives an impressive set of principles that should govern legitimate experimentation on humans.

1. Maximization of benefit and minimization of risk must be sought in every experiment (cf. pp. 527, 529, and 555).
2. Risk must be weighed in relation to potential medical gain rather than potential scientific gain (cf. pp. 476 and 555f).
3. All possible laboratory and animal tests must preceed experimentation on humans. He reports with horror the account of an investigator testing immunity on orphans: "Perhaps I should have made initial trials with animals. Calves would have been proper, but were too expensive to procure." (p. 535; cp. p. 555).
4. Subjects must give informed consent. Moll is quite aware of ways in which investigators, often unconsciously, coerce consent. He is, of course, also aware of outright deceit by investigators. As noted above, he knows of the suggestion that only physicians are able to give informed consent. Although he does not adopt that idea as his own, he does suggest that valid consent pre-supposes explanation of the total scope of the experiment and supposes that only well-educated persons possess the preparation essential to giving valid consent (cf. pp. 551 and 564ff).
5. Written consent should be obtained for relatively invasive procedures. He even supposes government regulation

might be required in this area. The rights of patients must be preserved even at the expense of some limits on research (cf. p. 568).

6. Informed consent does not in and of itself render an unduly risky experiment ethical, nor does consent justify the unnecessary repetition of experiments (cf. p. 557).
7. Particular groups are unsuitable as experimental subjects because of special vulnerabilities, e.g., the dying, those condemned to death by the State, various categories of institutionalized persons, and children (cf. pp. 515, 538, 541, and 531, respectively). The protection of the dying is a recurring concern (cf. pp. 525, 538, 544, and 547ff). He notes that even hangmen seek to diminish the cruelty of death by execution; physicians should do no less.
8. Experimental results must be truthfully reported (cf. pp. 571ff).
9. The results of research must be properly published, which is acknowledged to be a difficult task in the face of professional cliques and government monopoly in certain areas (cf. pp. 579ff).
10. Experiments must be conducted by competent persons with clearly defined responsibilities (cf. passim).

Although he thinks government action may be necessary on some fronts, he is opposed to the State becoming chiefly responsible for the investigation of new treatments. To the suggestion that a Federal Office be established to test all new remedies, he replies with vigorous objections. "For public institutions have in the past been not a bit more conscientious than private physicians in testing new healing remedies." (p. 598). Some of his most grisly examples concern experiments in state-supported health facilities. In the final analysis, the task of assuring ethical conduct in human experimentation must be performed by the profession.

If we compared virtually any phase of medical activity and organization in 1902 with its successor situation today, we would probably feel obliged to emphasize the enormous change that has occurred. How astonishing it is, then, to discover in this turn of the century treatment of experimental ethics most of the themes that still characterize the discussion today. When one compares Moll's analysis with the Tokyo revision of the Helsinki Declaration in 1975, there are very few major gaps in Moll's work. He does not know about later experimental methodologies and their special problems—the controlled clinical trial, the single- and double-blind experiments, and so on. Nor has he firmly worked out the distinction between therapeutic and nontherapeutic research, though it is implied in much that he says. And he gives no thought to experiments that intend to alter personality. But most of the rest is present.

How are we to explain this stability of moral perspective in the midst of so much change? Several possibilities suggest themselves.

1. Perhaps medical practice and research have changed less than is often alleged.
2. Perhaps we are in touch with values and norms that are central to human existnce despite all change at more superficial levels.
3. Perhaps the crucial turn is in the selection of the model in terms of which we will examine the question; while Bernard left us in some doubt about the status of experimental subjects, Moll clearly classifies them with patients, thereby joining their fate to a long-nourished set of norms and values.
4. Perhaps more attention should be paid to the carriers of various norms and values; the stability in experimental ethics may mainly reflect the stability of the profession that is the carrier of that ethic.
5. Perhaps the relevant professions have been unable or unwilling to respond to their genuinely new situation, because of some set of vested interests or lack of genius.

All of these hypotheses deserve study and reflection, though I am inclined to believe that 2, 3, and 4 are closer to the truth than are 1 and 5. Attention to 4 may be especially pressing since there are contemporary signs that centrifugal forces are powerfully at work in the medical profession in various parts of the world.

Our next landmark is also German in origin. It represents, so far as I know, the first attempt on the part of public authorities to regulate experimentation on humans apart from the very incomplete Prussian attempt referred to by Moll. However, the Guidelines were issued by the Reichsgesundheitsrat in 1931, and it would appear that the political upheavals of that period condemned them to a purely theoretical existence.[7] But they are fascinating in that form, and render all the more ironic the fact that German physicians were to become involved in the concentration camp experiments. Although I have been unable to follow in detail the development of the Guidelines, two drafts were published, the first under the title: Guidelines for the Administration of Scientific Trials on Humans.[8] The second title is enlarged in a way that reveals a significant distinction—Guidelines for Novel Therapies and for the Administration of Scientific Trials on Humans.[9] The change in titles makes clear that the authors of the Guidelines were interested in making a distinction between therapeutic and nontherapeutic research. Actually the first draft made the distinction in the body of the text; indeed the final draft introduces only minor modifications of the earlier text by way of rearranging, abbreviating, and strengthening various passages. The Guidelines are far closer to the Helsinki Declaration of 1964 than to the Nuremberg Code of 1947. The Guidelines offer what is primarily a social utility legitimation for human experimentation; a social need for progress in both therapy and science is asserted. But in order for these needs to be satisfied in an ethical manner, the following safeguards were to be observed.

1. Procedures must harmonize with medical ethics and the rules of the physician's art and science.
2. There must be a proper relation between anticipated utility and risks.
3. All relevant animal experiments must be done first.
4. Except in life-threatening situations, the subject or his legal representative must give unambiguous prior consent *(Einwilligung)* on the basis of a thorough explanation.
5. Special care must be exercised with subjects under 18 years of age.
6. The social need of potential subjects must never be exploited in order to recruit experimental subjects.
7. Special care must be used when microorganisms are used.
8. Supervising physicians must be directly responsible, or explicitly delegate their authority.
9. Records of all experiments must be kept and include information on the goals, justifications, procedures, and consent of the subjects or patient–subjects.
10. Published reults must maintain respect for the subjects.

The foregoing apply to all who are recruited for therapeutic research. When it is a purely scientific trial, the restrictions are somewhat tighter. Inability to give consent now eliminates the possibility of participation; children and youth under 18 are never to be used for such studies if there is the slightest danger;

such experiments on the dying are forbidden; and it must be impossible to make any further progress in the laboratory.

Many of the particulars are justified by reference to medical ethics *(die aerztliche Ethik).* It seems clear that the authors saw themselves as heirs to a professional ethic that was available for use by public regulators. Indeed, but for the distinction between therapeutic and nontherapeutic research, there is little here that cannot also be found in Moll. The frame of reference provided by the physician–patient relation still dominates, even being consciously extended to the realm of nontherapeutic research that is conducted by physicians. On the other hand, there are signs that the carrier group for this ethic is not wholly trusted. For it was envisaged that physicians in a wide variety of health care facilities would be required to sign a pledge to observe the Guidelines, and the Guidelines end with a paragraph urging vigorous attention to these matters in medical education.

Our next landmark is probably well known to many of us and can be treated more briefly. It might also be thought of as a product of the German experience, but that is a half truth at most. Although the Nuremberg Code was composed in Germany and represented a response to the German experience, it bears the marks of composition by Americans. Telford Taylor, a leading actor in the Nuremberg War Crimes Trials, has noted that the prosecutors in the medical part of the trials were embarrassed to discover that the German physicians and lawyers seemed to be in control of a much more detailed and refined body of lore concerning the ethics of human experiments than were the prosecution experts.[10] This should hardly surprise us in view of our review of Moll and the 1931 Guidelines.

It appears that the prosecutors deemed it wiser to establish a new Code than to avail themselves of the already established Guidelines. Whatever their reasons may have been, a Code was produced that shifted attention to the issue of free and informed consent to a degree that cannot be duplicated in earlier discussions of which I am aware. Thus the Code's longest and first section begins: "The voluntary consent of the human subject is absolutely essential." No other requirement is stated in absolute terms. One explanation of this focus on consent might be that it is written in response to the revelations of the concentration camp experience in which consent was so obviously lacking. However, many other traditional canons of experimental medicine were also violated in the concentration camps; the post-Nazi German discussion was as interested in these other defects as it was in the consent problem. My own hypothesis is that the preoccupation with consent derives, rather, from an American cultural tendency. It is at least noteworthy that the American discussion of experimental ethics since World War II has been dominated by the consent issue. To point out but one symptom of this, what is probably the most widely used American bibliography in medical ethics entitles the relevant section—"Experimentation and Consent."[11]

There are, of course, many other provisions in the Nuremberg Code, but very little is added to what we have found in earlier statements. The most original contribution of the Code is its emphasis on free, informed consent as the *sine qua non* of ethical experimentation on humans.

It would no doubt be worth our time reviewing still more recent developments in the ethics of human experimentation, especially the much cited Helsinki Declaration. But I will refrain from doing so for three reasons: many people are familiar with the Helsinki Declaration; it backs away from the Nuremberg Code's stress on consent, and renews the tradition that we have found to prevail earlier in the century; and as the resident ethics "historian" for the purposes of this work, the Helsinki Declaration seems to fall outside my sphere of responsibility as a relatively recent event.

Two observations about the Declaration are, however, in order. 1. The notion of proxy consent, which in older documents had usually been approved primarily in connection with life-threatening situations and the status of children, is further unpacked so as to make explicit allowance for any kind of "physical or mental incapacity" (paragraph 11 of the Basic Principles). This does not strike me as a great new departure, but it does unambiguously include the population of interest to this conference. 2. Reference is also made to a duty "to minimize the impact of the study on the subject's physical and mental integrity and on the personality of the subject," (paragraph 6 of the Basic Principles). This provision may often have specific relation to experiments with the population group under discussion. Though the requirement probably was not intended to eliminate experimentation intended to "improve" mental integrity and personality, it reminds us that what is and what is not "improvement" is often open to controversy, and that special restraint is called for in this area.

Alzheimer's Dementia: Contemporary Tensions Between Clinical Needs and Moral Boundaries

Let us turn to some possible implications of the tradition we have sketched with respect to research done with the population that is of central interest to this conference.

Clearly the tradition we have outlined yields what has been identified as the chief problem in the clinical treatment of Alzheimer's disease patients. Those suffering from Senile Dementia do not seem to be persons capable of competent consent, at least during later phases of the disease. Were not informed consent a canon for ethical research, there would be few barriers to research on this population. If, for instance, a military model had been selected for our understanding of research subjects, we might well imagine certain populations being ordered to participate in certain research studies. That might well include persons with this disease, possibly with the comfort that they would not suffer the additional burden of fully understanding what pains and risks they would confront. A social utility legitimation for experimentation on humans free from the consent requirement could make this population a prime candidate for certain kinds of experimentation. But the tradition is built on the physician–patient treatment model and has insisted on the consent requirement, which has no doubt been a barrier to both therapeutic and nontherapeutic research. The consent requirement has prevented exploitive experimentation, but it has also probably excluded experimentation that might have produced benefits.

A proposed solution to this apparent dilemma has been to shift consent responsibilities to guardians, to look for possibilities of proxy consent. But it is not clear that this will always yield a moral solution, since guardians as assigned by custom or law may not always in fact be devoted to the best interests of the potential subject. It seems to me this danger is especially present in the American context, in which consent has not only been viewed as a necessary condition for ethical experiments, but sometimes viewed as a sufficient condition. More than once I have heard persons associated with medical research respond to a proposal by saying that though they could not imagine participating in such an experiment, they supposed it would be alright for anyone who would sign the consent form. The longer tradition to which I have pointed your attention was always sensitive to the need to weigh consent along with a number of other considerations. Insofar as research proposals have passed through our current review processes, attention will presumably have been paid to these other matters before any sort of proxy consent would be sought. But one should keep in mind the empirical evidence that even review committees seem to focus much of their attention on consent procedures.[12]

It is tempting to suppose that there may be parallels between this population and other populations that present consent problems. Solutions proposed for these populations might be applied here. There have been, e.g., suggestions that children might be volunteered for nontherapeutic research on various grounds such as duty to society, guardian ratification of assent that falls short of competent consent, the admittedly remote possibility that any childhood disease might attack a child whether he is presently afflicted or not, and so on. In my opinion, even if these arguments persuade one regarding children, they will not be easily applied to senile persons who are usually closer to the other end of their biographies. In at least some sense, these people have done their duty to society; they may be as incapable of assent as they are of consent; they are already involved in a serious disease condition. Other consent–incompetent populations are often as difficult to include in ethical experimentation as Alzheimer's patients, or they possess contrasting characteristics that make analogies slippery at best.

The inherited tradition, in my opinion, offers very restrictive possibilities for research with this group. It certainly allows therapeutic research with proxy consent if other criteria for an ethical experiment have been satisfied, but it would seem to exclude all nontherapeutic research. The traditional ethic has long shown a special concern for the dying, to protect them from experimental exploitation. The group under discussion is presumably not usually on the terminal clinical list, but it presents many of the same temptations and vulnerabilities that belong to the dying.

If this strikes some as being overly restrictive, there remain several options. One can, of course, seek to overturn the traditional ethic we have described. Clearly many attempts of this sort have been made in this century in various parts of the world, most notably under conditions of real or alleged national emergency. Were I disposed to do this, I would attack as most vulnerable the application of the physician–patient model to various experimental situations. The military model is one possible alternative, as is the business contract model, and one can imagine others. All of these would, of course, present their own unique difficulties when applied to the population in question.

Another approach may seem still more bizarre to many. Without trying to tell the whole story here, it is my opinion that a tremendous "socialization" has occurred on the side of experimentors during the 20th century. Moll visualized an innovative individual physician asking an individual patient to participate in an experiment. That is still a picture that can be drawn at the moment when consent is requested. But behind the physician we will often find a large and complex research team, a highly elaborated hospital structure, powerful financial sponsors, institutional review boards, government agencies, and so on. Though much of this apparatus is designed to protect patients, the structure as a whole has a vested interest in the conduct of medical research. The patient or subject has little by way of countervailing power.

Among ways in which patients and subjects might "socialize" their side of the equation would be to organize prospective consideration of participation in future research. In modern society, it has become rather likely that we will at some point in our lives be desired as research subjects; we should prepare for that eventuality as we prepare for other probable experiences. It is morally defective that so many persons are first invited to consider participation in research only in the condition of patient vulnerability. Perhaps there could be something analogous to the so-called "living will" in which still-competent persons would give some indication of their openness to participation in research should they become incompetent. We have a forerunner in the growing use of driver's license notations regarding the use of our bodies if we are killed. In my preliminary thinking about such a document, the problems seem enormous, but less large than those surrounding what we currently call the "living will." There would, of course, have to be some specification of acceptable and non-acceptable research areas. This practice would at least address the fact that persons carry whole biographies into experimental situations; that would be a moral gain over many experimental situations in which subjects are treated as current events that have volunteered.

Let my last word, however, be a word of appreciation for the historical tradition we have outlined respecting the ethics of experimentation on humans. It is often inelegant and incomplete when viewed from various moral perspectives; it has not and will not prevent all unethical experiments even as it defines them; it clearly pays some cost in terms of knowledge foregone or postponed. But it has made possible a long and continuing relation between modern science and medical care in a manner that, in my opinion, deserves continued existence. It is a marriage that could easily have failed.

Notes and References

[1]Viktor von Weizsaecker, "Euthanasie and Menschenversuche," special reprint from *Psyche,* I, 1947, Heidelberg: Verlag Lambert Schneider; and Hans Jonas, "Philosophical Reflections on Experimenting with Human Subjects," in Paul Freund, ed. *Daedalus,* Spring 1969, 219–247.

[2]Cf., e.g., the essay of Benedict Ashley, "Ethics of Experimenting with Persons," in J. Schoolar and C. M. Gaitz, *Research and the Psychiatric Patient* (New York: Brunner/Mazel, 1975), esp. p. 17.

[3]Albert Moll, *Aerztliche Ethik* (Stuttgart: Enke Verlag, 1902), pp. 565f.

[4]Irving Ladimer and R. W. Newman, *Clinical Investigations in Medicine: Legal, Ethical and Moral Aspects* (Boston University Law and Medicine Research Institute, 1963), p. 1.

[5]The ideas and quotations in this paragraph all come from the translation of the *Introduction . . .* by Henry Greene, Henry Schuman, Inc., 1949 ed. pp. 99–105. These were the principles of a very self-assured personality. Thus he states in his *Cahier Rouge*: "The strength of men who keep thinking about the same thing. It is not in writing easily or in acting easily. It is to have an awareness of what is good—I have it. It always comes to this. This is the essential thing . . . Perfection is possible. This epoch will bear my mark." In F. Grande and M. Visscher, *Claude Bernard and Experimental Medicine* (Cambridge: Schenkman Publishing 1967), second paganation, p. 37.

[6]Moll, *op. cit.*, p. 5. The page numbers of subsequent quotations will be given in the text.

[7]One contemporary jurist, noting that the Guidelines probably lost all legal force with the end of World War II, has suggested that they be readopted. Cf. H. J. Wagner, "Heilsversuche und Experimente aus Rechtsmedizinischensicht," reprint of lecture given January 8, 1975.

[8]*Reichsgesundheitsblatt*, 1931, p. 531.

[9]*Reichsgesundheitsblatt*, 1931, pp. 174 f.

[10]*The Hastings Center Report*, 6 (4), Aug. 1976, Special Supplement, pp. 6ff.

[11]*Bibliography of Society, Ethics and the Life Sciences 1979–80*, Hastings-on-Hudson, The Hastings Center, 1978, p. 30.

[12]Cf., e.g., Bradford Gray, "Complexities of Informed Consent," *Annals of the AAPS*, 437, May 1978, 43.

Research on Senile Dementia of the Alzheimer's Type:

Ethical Issues Involving Informed Consent

Christine Cassel

Introduction

There are certain values that we would all, in general, agree upon and therefore designate as prevailing cultural or professional norms. These include integrity (or honesty), respect for other persons, justice (or fairness), and compassion. The issue we focus on here is troublesome because it seems to embody conflicting values. The conflict is at times more apparent than real, and one function of philosophical analysis is to clarify issues in order to resolve conflict. There are also, however, conflicts that are and will remain very real. We are asking difficult questions that need to be asked. The asking is as important as the answer, so we should give these questions a full measure of our attention rather than only seek the route that seems most expeditious. If any light can be shed on such engimas of value and meaning, it is likely to occur in an interdisciplinary discourse such as this. Since we are all in some way limited by the value hierarchies of our individual professions, we only stand to be enlightened by a convergence of various relevant perspectives.

Some of the kinds of questions underlying our task at this conference are as follows: Is autonomy the most significant human attribute? Is paternalism the solid basis for good medical practice, or is it an egoistic delusion of the profession? What is the value of biomedical research to human society? What sacrifices, if any, of other values are warranted for the sake of increasing knowledge? Is there a special respect due to the brain as the symbolic seat of personhood? What are the origin and nature of our obligations to others in the human community—present and future?

These questions and others like them must inform our approaches to the concrete focus of informed consent in research on SDAT. In this sense, addressing the problem of consent requires and enables us to take a hard look at the most basic values of our relationships to those who suffer this disease, their communities, our entire society and indeed, the very meaning of the work we do.

Some of the basic groundwork has already been done. The National Commission for the Protection of Human Subjects of Biomedical and Behavioral Research published in 1978 its Report and Recommendations on Research Involving Those Institutionalized as Mentally Infirm.[1] The term "institutionalized as mentally infirm" is defined to include those who are mentally ill, mentally retarded, emotionally disturbed, psychotic, or senile, as well as those whose impairments are "similar" and who therefore reside in "an institution." These recommendations are carefully considered guidelines intended to protect the rights and well being of the subjects without unduly impeding important research that might significantly benefit them or others like them. Institutionalization in itself causes distinct problems in ensuring of voluntariness and, when chronic, can affect the level of mental and social functioning. Thus the commission saw fit to treat this group as a whole, and their recommendations are in many ways similar to those made for other institutionalized groups, such as prisoners.

The area of overlap between our subject in this conference and this report of the National Commission is the group referred to as "senile." The majority of elderly persons in nursing homes are said to have some form of "senility,"[2] but it is unknown how many of these actually have SDAT. By far a greater number of persons afflicted with SDAT are not institutionalized. Although estimates are that approximately 1 million persons suffer from this disease, there may actually be many more. What few epidemiological studies there are have been confounded by the probable high incidence of misdiagnosis and the reluctance of families to bring the problem to physicians in its early stages. There is no doubt, however, that an effective treatment, cure, or preventive strategy could prolong useful life and save enormous suffering.

When we address the ethical issues in SDAT research, the patient group is extremely varied in terms of location, social circumstance, and severity of disease. Some may have other diseases that account for their dementia. For this reason, brain biopsy has been recommended as a way of improving the clarity of research data. Considering the use of an invasive, nontherapeutic procedure such as this prompts a closer look at the general problem of informed consent in this population.

The ritual of informed consent reminds us that persons who happen to be patients or research subjects are not means to an end, but rather partners in an enterprise that they join without deception or coercion. This ideal relationship is called into question in every respect when the patient is cognitively impaired.

Although it is important to remember that therapeutic and nontherapeutic research present some different ethical considerations, a patient may consent to either if he or she is considered competent. The consent process can best be studied by dividing SDAT subjects into two groups, each presenting different kinds of problems. The first group are those in whom the disease has progressed so far that they are globally incompetent to make decisions on their own behalf. In this group the ethical problem centers on the validity of various forms of proxy consent. The second difficult group are those in whom competence is borderline or fluctuating. We have no clearly correct methodology for establishing competence (or the lack of it) in this group, nor is there a clear philosophical basis for establishing the threshold of such a concern.

Problems of Evaluating Competence

There have been many studies demonstrating that subjects do not understand, or at least do not retain, most of the information given them. This is often attributed to the highly technical language used by the researcher, and the subjects' unwillingness to press for a simpler explanation. The subject may feel intimidated by the greater knowledge and power of the physician, and thus be unwilling to press for clarification. These studies have all been done with patients or subjects of ostensibly normal mental capacity.

In persons with mild to moderate dementia, the same problems exist, but to a greater degree. Still, knowing that many normal subjects do not understand or remember what they have been told calls into question any arbitrary designation of competence based on ability to remember. It seems clear to me, as a clinician, that there are patients who have significant short-term memory deficits that do not impair their ability to attend to the information that is being given to them and to understand it at the time it is being given. If they do not remember it the next day, does this invalidate their consent? I think it does not. Others may disagree, but such disagreement means we cannot use the simple mechanism of a recall test to evaluate competence.

Since it is so difficult to know how to assess competence by using a "process" measure, i.e., how much was understood, some writers have suggested using an "outcome" measure, i.e., is the patients decision reasonable in the context of their life situation? Although this criterion of reasonableness is one that has often been brought up in de jure competence decisions, it seems to be an extremely tenuous criterion to use because it is so subjective. Roth et al. have pointed out a similar phenomenon in the situation of informed consent to treatment in psychiatric patients.[3] The patient's competence to consent is rarely called into question unless the patient makes a decision in conflict with the physicians' assessment of what is medically indicated for the patient. The patient's decision may then be considered "irrational." "Reasonableness" or "rationality" are rarely valid as sole criteria for the judgment of competence because medical decisions are often value judgments like any others. As medical technology has advanced, the risks as well as the benefits of interventions have increased. What the physician sees as an acceptable risk depends largely on his or her own value system, which may differ significantly from that of the patient. McNeil and her group have published two studies demonstrating that patients as a group differ markedly in their attitudes toward risk from physicians as a group.[4,5] There are also significant intragroup variations among the patients.

The confusion in this assessment of competence has two components: (1) the assumption that a "rational" decision equals a "competent" decision and (2) that a rational decision is to be judged by the value system of physician, especially since physicians value rationality so highly. Rationality may be more of a value judgement than a measure of competence. Case law has supported patients rights to make irrational decisions.[6]

"Reasonableness" criteria has been used in situations of evaluating competence to consent to treatment, not to experimentation. We might, in certain cases, support such a measure of competence when the patient's well-being is at stake facing a medical or surgical illness, when our own conscience suffers at the betrayal of an imperative of benevolent paternalism. But participation in research is more optional, and therefore we have no overriding paternalistic duty pressing us to question the "reasonableness" of the patient's decision.

Conflict of Interest

Voluntariness must be at the core of the consent process, and is in fact the attribute which that process is designed to ensure. Thus, freedom from coercion is an essential component of valid informed consent. If the researcher is also the patient's personal physician, there may be a subtle but powerful pressure on the patient or proxy to consent to participate in a clinical trial. He may want to please his doctor, or at least not want to appear to disagree. It is also easy for the physician to describe the experimental protocol in a way that emphasizes its promise of benefit to the patient and de-emphasises the risk. There is a tendency in medical practice for a "new treatment" to be equated with a "better treatment," even though the premise of a clinical trial is that the question has not yet been answered.

The physician–researcher is likewise under subtle pressures that may make it difficult for him to present the choice to the patient in an entirely unbiased fashion. He may really believe that the treatment will help. He may need to recruit a certain number of patients in order to get valid statistics to interpret the results of the trial. He may need to finish this research project in order to be promoted. Some of these motivations are open to more criticism than others, but all can be unconscious influences on the informed consent process that result in diminution of the patient's freedom in choosing. In persons with SDAT, personality is more frail, trust is harder to come by, and once established, may be easier to abuse.

For these reasons many writers have suggested that an investigator should never wear both hats of researcher and primary physician. The dangers of conflict of interest are possibly too great. It needs to be mentioned here, however, that a trusting relationship with any physician is an immensely valuable thing for a patient with SDAT, and therefore an absolute prohibition against the physician playing both those roles may finally not be in the patient's best interest. An optimal clinical setting for an SDAT patient includes consistent relationships with a relatively small number of caregivers. Therefore, the moral character of investigators is a crucial aspect of the ethical conduct of research.[7,8] Devising a test of ethical competence for investigators may be as relevant as trying to find an adequate test of mental competence for patient–subjects.

Strategies that would allow the physician–researcher to be one and the same person, while minimizing the coercive effect during the consent procedure, include designating another professional to monitor the process. This person could be a patient advocate or consent auditor, as recommended by the National Commission. We must remember again, however, that in SDAT the introduction of a new and strange person to the consent process would surely be confusing to the subjects. Such persons would have to be very carefully trained and turnover kept to a minimum.

Respect for Autonomy vs Paternalism

At the core of our discomfort about the process of informed consent with SDAT patients is the tension between two basic values. On the one hand, we want to respect the patient's autonomy, and therefore feel obligated to do whatever we can to enhance that autonomy. On the other hand, as physicians and health care professionals, we want to do what is best for the patient.

George Alexander has said, "It is important to recognize that however benevolent the intention of those who would seek to substitute other decision makers for the aged, persons deprived of the right to decide for themselves will have lost the most basic attribute of citizenship."[9] Once a legal judgment of incompetence is made, it is not easy to have it reversed. Once a label of incompetence has been used in a medical chart, its presence there will continue to influence the attitudes of all health professionals who have access to the chart. This can be a devastating blow to the level of respect a patient receives during contact with the health care system. These arguments will support a tendency to err on the side of respecting autonomy, perhaps stretching our definitions of competence, and considering proxy consent only in the extreme, perhaps when the patient is unable to communicate at all.

I would also argue that there is something to be said for paternalism. Thomas Halper has poignantly asked, "Of what value is liberty to one who is buffetted by the violent winds of the mind?"[10] Perhaps our greatest respect for these patients has to include some element of paternalism. In our society, and especially in our medical institutions, aged persons with some degree of mental impairment may need, and therefore deserve, some special protection against exploitation and neglect. It may be easy enough to allow cognitively impaired subjects to agree to investigative procedures in the name of respecting their autonomy, if that agreement furthers our own projects. We must maintain a constant vigilance (or index of suspicion) that the subject is possibly not acting in the truest sense of autonomy and "being buffetted by the violent winds of the mind" needs therefore our most careful protection.

Paternalism, Justice, and Quality of Care

In discussing the problem of paternalism as it relates to the mentally retarded (another group whose cognitive deficits tend to obscure respect for the affective aspect of their humanness), Dan Wikler has described the tension between the values of social welfare (and paternalism as a mechanism for enforcing these values) and equal liberty for all (with its attendant risks of abandonment of the needy). "Given our concern that the mildly retarded not be pushed out onto a dangerous world in which they may come to ruin we have two choices. We may change the world so as to render it safer for all. Or we may simply refrain from allowing the retarded access to it."[11] Thus, he reduces the dilemma of paternalism to a question about our social and institutional responsibilities. In the context of research with SDAT, there is an obvious parallel. If we can construct a therapeutic and research context that is truly in the best interest of this group of patients, then measures to protect them from us not be so extensive. Nor would we need to forego our basic obligation of caring. Limiting research funding to only those nursing homes and hospitals that provide compassionate care and optimal environments for the demented will reduce their vulnerability and thereby allow for more ethical conduct of research. The subjects' participation must not, however, become a condition that is prerequisite for access to quality of care. This is a form of coercion that is powerful and insidious, as demonstrated by the notorious Willowbrook studies on hepatitis vaccine.[12]

It is important to point out that encouraging therapeutic research in SDAT can have many benefits for the patients. The obvious one is the possible improvement in their abilities to function meaningfully in a social context. Stanley Hauerwas has written eloquently of the need to include the chronically ill in the human community as a way of giving meaning to their suffering.[13] Actively engaging with them in any shared endeavor could go a long way towards this goal.

Whatever mechanisms of protection of patients rights are adopted, they should not needlessly reinforce the sense of "otherness" of the elderly demented. Making a special category of persons has a tendency to separate them from "normals." Such a move may enhance the behaviors of respect, but actually erode the true nature of mutual respect for a fellow human, and is certainly dangerous for the complimentary values of compassion and justice.

It has often been observed that elderly patients with dementia are classic examples of "uninteresting patients," and as such, they tend to be neglected and undervalued in our teaching hospitals.[14] This is a critical failure because the training of physicians occurs in this setting, and negative attitudes toward the elderly are reinforced. If faculty are actively interested in the problems of this group, and if exciting research is taking place in the teaching hospitals or nursing homes, then these patients have a chance to become "interesting." I have observed this shift of attitudes around the problem of incontinence when a research project in that area was initiated.

Thus, if the criteria for humane care are met, participation in research may be a benefit to the subjects as well as to future sufferers of the disease. Researchers can then consider their work of present value to the subjects as well a as future social value.

Ethics and Methodology

Full consideration of the social value of research goes beyond the ethical implications of the sites where investigators are conducted and the quality of patient care. It also includes the setting of priorities in the kinds of research to be encouraged, and problems of methodology. Behavioral, social, and health care systems research may have more chance for an immediate benefit to the subjects. Strictly biomedical research stands to benefit future populations. The latter forms have been more favored in medical contexts because they are more likely to result in data that can be quantified and carefully controlled. The methodologies of behavioral and social science research are inherently less precise, and therefore less satisfying to the medically trained critic. This does not, however, necessarily mean that their results are less important. We must carefully balance the values of scientific rigor against the tremendous claim of the subjects to respectful treatment. Whatever research we do ought to be as rigorous as possible—ought to be *good* science. Ethical constraints may, however, limit the degree of precision that can be reached. But at our present degree of ignorance about this disease, sophisticated statistical methods will still allow meaningful interpretation of less than perfect data. In clinical medicine we are taught the necessity of working constantly with ambiguity and uncertainty. In evaluating research in SDAT, we likewise may have to tolerate some ambiguity and be more creative in the kinds of questions we ask, and the methods we invent to answer them.

Conclusion

The ethical obligation to proceed—and to proceed vigorously—with SDAT research derives from the large numbers of persons afflicted; the social significance of that affliction and the suffering it causes to patients, families, and communities; and some retribution owed because of the medical profession's neglect of this problem for so many years. But ethical caution is supported by considerations of respect for persons and the basic obligation of physicians to act in the best interest of their individual patients.

Informed consent is a ritual that formalizes respect for the person as an individual. Its constraints may present limits that seem to impede the progress of research, but should help us continue work in a way that is more basically beneficial to the least advantaged persons in the research community—the patients themselves. In considerations ranging from determination of competency and proxy consent to the setting of priorities for research questions, flexibility and caution must go hand in hand.

Notes and References

[1]The National Commission for the Protection of Human Subjects of Biomedical and Behavioral Research. *Report and Recommendations: Research Involving Those Institutionalized as Mentally Infirm,* DHEW Publication No. (05) 78–0006, 1978.

[2]NIA Task Force. "Senility Reconsidered: Treatment Possibilities for Mental Impairment in the Elderly," *Journal of the American Medical Association* 244, 1980, 259–263.

[3]L. Roth, A. Meisel, and C. Lidz, "Tests of Competency to Consent to Treatment," *American Journal of Psychiatry* 134, 1977, 281–285.

[4]B. J. McNeil, R. Weichselbaum, and S. G. Pauker, "Fallacy of the Five Year Survival in Lung Cancer," *New England Journal of Medicine* 299, 1978, 1397–1401.

[5]B. J. McNeil, R. Weichselbaum, and S. G. Pauker, "Speech and Survival: Tradeoffs between Quality and Quantity of Life in Laryrgeal Cancer," *New England Journal of Medicine* 305, 1981, 982–987.

[6]G. Annas, *The Rights of Hospital Patients: The Basic ACLU Guide to a Hospital Patient's Rights* (New York: Discus Books, 1975).

[7]H. K. Beecher, "Ethics and Clinical Research," *New England Journal of Medicine* 274, 1966, 1354–1368.

[8]B. Barber, "The Ethics of Experimentation with Human Beings," *Scientific American* 234, 1976, 25–31.

[9]G. Alexander and T. H. D. Lewin, *The Aged and the Need for Surrogate Management* (Syracuse: Syracuse University Press, 1972).

[10]T. Halper, "The Double-Edged Sword: Paternalism as a Policy in the Problems of Aging," *Milbank Memorial Fund Quarterly* 58, 1980, 472–499.

[11]D. Wikler, "Paternalism and the Mildly Retarded," *Philosophy and Public Affairs* 8, 1979, 392.

[12]S. Krugman, "Experiments at Willowbrook State School," *Lancet* (1971): 966–967.

[13]S. Hawerwas, "Reflections on Suffering, Death and Medicine," *Ethics in Science and Medicine* 6, 1979, 229–237.

[14]C. Cassel and A. Jameton, "Dementia in the Elderly: An Analysis of Medical Responsibility," *Annals of Internal Medicine* 84, 1981, 802–807.

An Alternative Approach to Informed Consent in Research with Vulnerable Patients

Andrew Jameton

Introduction

The present system of regulating clinical trials with close attention to the ethical aspects of research presupposes that research can create a conflict of interest for physicians and researchers. In the energetic pursuit of a new idea or possible treatment, researchers may be tempted to set aside patients' immediate welfare for speculative benefits to future patients. Reluctantly, we admit that much research is of little value to human welfare, that some research is likely to harm patients, and even that scientists sometimes act in untrustworthy ways.[1] In order to protect patients from harm and researchers from temptation, we approach research on human subjects cautiously.

The Dilemma

When we turn this cautious eye to the problem of conducting research on patients with Senile Dementia of the Alzheimer's Type (SDAT), we experience considerable ethical discomfort. On one hand, SDAT is a serious health problem, and research could result in therapies improving human welfare.[2–7] On the other hand, patients with advanced SDAT are among those about whom we feel the most hesitation in doing research. This is because voluntary participation, informed consent, mutual understanding, and respect are at the core of our present system of protecting patients from potentially ambitious inquiries. Many SDAT patients have diminished or uncertain competence to consent, live under conditions of extreme dependency, cannot exert voluntary control over much of their conduct or many bodily functions, and suffer from gloomy or discouraging prognoses that many physicians find uninteresting. In addition, such patients are usually considered unappealing by virtue of discrimination against elders, the disabling consequences of their disease, and sometimes, poverty.[8–10] Often, patients with SDAT are unable to speak for or defend their own interests or attract reliable allies and advocates.[11,12]

This is, of course, not true of all SDAT patients. Not all people with SDAT are institutionalized, and in its early stages, the disease does not affect mental competence. Even where mental competence is diminished, specific disabilities can be compensated for in order to maximize voluntary participation. For example, forgetfulness is less important when the patient is given frequent reminders. Problems in reasoning can be diminished by careful assistance with thinking through an issue. Agitated patients can be calmed and reassured. Where studies present little or no risk—such as with many innovations in modes of patient management—a permissive standard for consent might be all that we require. Surely, for some SDAT patients the problems of informed consent and patient protection are not pressing and can be solved by careful attention to the specific disabilities of each patient.

Yet, there remain many SDAT patients who are so severely demented that informed consent is impossible. Even if such patients were to agree to participate in a clinical trial, one could not rely on their appreciating the meaning of that agreement. This situation becomes a dilemma when the only group on which certain types of studies can be done is severely demented patients. For example, a study may require samples of SDAT damaged brain tissue that is most likely to be found in advanced cases of the disease.[13,14] The only reasonable mode of obtaining tissue is via brain biopsy, a procedure that, although not very risky, is nontherapeutic, invasive, and involves more than minimal risk.[13–17]

Should we take the route chosen by Paul Ramsey with regard to the ethics of nontherapeutic research on children and virtually exclude severely demented patients from studies involving more than minimal risk?[18] Surely, if any group is to be excluded from experimental study by our present standards for ethically acceptable research, it is the institutionalized mentally infirm. The ethical argument for this position is that people should not be used merely as a means to an end, even a medically valid one. And, where competence and liberty are sufficiently diminished, a person can only be acted upon and cannot be treated as a colleague sharing in the research enterprise. Thus, in order to avoid violating principles of informed consent and voluntariness, such research would have to be avoided. Or, is there a way that research can be conducted ethically on these most unlikely subjects?

The Pressure for Research

We need to consider how pressing the proposed research is before exposing patients to risk.[4, 19–23] But this consideration raises as many questions as it answers. The value of a particular line of research is a realm that human subjects committees are loath to discuss. Most research is conducted in a complex context of research projects, is of indirect and uncertain benefit to future patients, and within a realm close to the autonomy of researchers. It is hard enough to make judgments with regard to the value of research on subjects who are fully competent. Must research be more pressing than usual in order to justify exposing vulnerable subjects, such as SDAT patients, to the same risks as healthy subjects? How relevant is it that SDAT patients represent the same group of people most likely to be benefited by therapeutic results of such research?

In regard to invasive physiological studies, one could argue that the social context of SDAT has a profound effect on the suffering associated with it. The process of deinstitutionalization combined with the restricted availability of resources for those patients remaining in institutions cause as much misery as the disease itself. If such social problems are a major cause of suffering in SDAT, what then is the relevance of physiological approaches to the problem?

Perhaps, since the neurophysiological impact of SDAT on mental life is so devastating, it is overdrawing the significance of its social dimensions to give them such prominence here. Yet, before we even consider invasive modes of research on SDAT patients, we must be clear about the etiology of our perception of the need for it. We have the option of holding back on the more invasive forms of research. If our social priorities in SDAT research are more pressing than our medical priorities, a firm stand that invasive research *cannot* be done ethically may stimulate us to greater technological ingenuity in finding less invasive modes of study, more appropriate animal models, and so on. This would not be the first time that a moral problem was resolved, or at least averted, by technological means.

Stepping Back from Voluntariness

Suppose that after consideration, we conclude that it is vitally important to proceed even with invasive modes of SDAT research. Now we must exercise our moral ingenuity. Is there a way out of the dilemma that SDAT patients pose? Two types of resolution seem possible: We can carefully study how to apply the concepts of informed consent and voluntariness in extended or derivative forms to SDAT subjects in such a way that we can feel confident that they express respect. Or, we can step back from voluntariness and informed consent and find other means of protecting subjects. Both courses, I believe, deserve careful attention. Since so much research on ethics has been devoted to developing the concepts of informed consent and autonomy,[1,24–30] I would like to discuss briefly what would be required if we diminished the importance of voluntariness as a condition of participation in research.

To take this course, we need to understand why voluntariness and autonomy are so important to the protection of human subjects. If we can see what voluntariness is for, and if we can obtain the protections voluntariness provides without it, then we can think through the ethics of research without appealing to this difficult, metaphysical concept.

We need voluntariness and consent for two reasons. First, we live in an unsafe world. In a safer world, competence and voluntariness would not be so necessary. Second, patients often lack power in relationship to health professionals. When a powerful person looks out for the interests of a less powerful one, this paternalism, however well meaning, is likely to go awry. The person purportedly benefiting in such an arrangement is likely to be used as a means to the ends of the more powerful party. We can abandon the ethics of voluntary participation only if we are willing to prevent harm to subjects threatened by an unsafe world and inequalities of power.

Stepping back from voluntariness carries a considerable price. It requires modification or careful selection of the institutions in which we want to conduct research so that they are safe for all patients and so that patients share power with their caretakers.

Exemplary Care

The usual modes of protection include:

1. Careful selection of subjects in order to avoid injustice, involuntariness, manipulation, ignorance, discrimination, and so on
2. Limits on the types of research that can be done
3. Properly qualified researchers
4. Procedural guarantees, such as consent forms, review committees, and so on.[27–30]

For these conditions to be fulfilled we must rely on the sincere concern of care-giving and research personnel, and their appreciation of respect for patients. Ethics as a mode of regulation functions best when a set of principles is genuinely held by a group of people. Ethical principles are created to avoid systems of review. In order to provide the standard protections of human subjects, we must ultimately rely on everyone's good will.

The best evidence of the good will of providers is exemplary patient care, as demonstrated by the most ordinary and concrete terms: are patients in the institution well and happy? Do they receive excellent basic care and are they engaged in a variety of activities? Do families participate actively in the care of patients? Does the staff show good morale? In a cautious approach to SDAT research, we should make sure that research is done only in institutional settings where the very best care is provided.

I do not appeal to exemplary care only in order to ensure that the usual protections are carried out. Once the condition of exemplary care is met, along with the additional condition of equal power discussed below, they tend to replace the usual means by which patients are protected, because they are significant guarantees that ordinary regulation succeeds in any case. When we, as patients, live under good conditions among people who care about us and are cared for with perceptiveness, we are not likely to be asked to do things that are dangerous or irrelevant to our interests.

However, the good character of staff and researchers—their freedom from any wish to harm or exploit patients—is not adequate protection. This is because ineffective goodwill is possible, when in spite of the good intentions of staff, patient care is not adequate. Exemplary care is needed to show that an ethos adequate to the protection of vulnerable subjects has been established and is successful. Thus, in deciding whether to support a study, it is important to look not just at the quality of the protocol and the consent forms, but also to see how they fit into the system of patient care around them and what the quality of that patient care is.

Collective Patient Control of Care

Exemplary care is not enough. We must also consider the more difficult issue of patients' power, or lack thereof. What I call the "King Lear" problem illustrates this concern. At what point in King Lear's progressive dementia would you approach him or his family with questions about participating in a study? Consider, for example, a protocol that requires a prompt autopsy. In order to obtain his consent while he was still mentally competent, would you ask Lear about it when he makes his first errors of judgment, or later, when he is raging on the moor? Would you wait until he is in prison, or until he is grieving over Cordelia?

There is something splendidly irrelevant about conducting research in the face of the tragic and inexorable decline of our mental powers or those of our loved ones. This problem cannot be resolved by better use of informed consent during a tragic decline. Neither is it resolved by making more subtle distinctions among levels of voluntariness and competence. The only comfortable solution places the locus of decision-making in another time and place altogether. Suppose that during his reign, Lear accepted the advice of his court scientists and ordered prompt autopsies on all patients who died with symptoms of dementia. Then there would have been no need to approach him with the question of his own autopsy, and yet it would be a decision to which he made an important contribution. Indeed, Lear could reflect ruefully on the moor:

To the physicians my brain must go
To toy with for their protocols . . .

For patients in an institution, the move from being off a protocol to being on a protocol need not be a conscious decision. It can be part of the background of life, one of the givens of certain circumstances. Some of our grandest experiments—public education, the market system, and civilization itself—are given to us in the backgrounds of our lives. We participate in these experiments involuntarily, without informed consent, and yet we participate in them actively. But to conduct experimentation in this way, i.e., without informed consent or voluntariness, we must have collective substitutes for individual consent.

Our grander institutions can only appeal tenuously to a theoretical social contract to justify rights and obligations. In contrast, institutions for patient care provide the opportunity to create a social contract in fact. If institutions for patient care are controlled by everyone involved in them—especially patients and their families—and not just owners, administrators, and professionals, then we could approach a collective substitute for voluntariness. Thus, I am calling for limiting research to institutions in which there is greater control by patients and their families, not just at the point of consuming health care services, but in the organization and management of the institutions themselves.

I am not suggesting that demented patients should administer institutions. In any collective substitute for individual consent, variations in competence must still be taken into account. Instead, I am advocating active participation by competent patients and their families in creating and managing care-giving institutions. When responsibility is shared broadly by those affected by an institution, there is less fertile soil for paternalism. There are two reasons for this. First, since everyone is relatively equal in power, there is less opportunity for some to dominate others. Acts that are paternalistic between persons unequal in power, merely express mutual support among equals. Second, in settings where equality prevails, thinking about the good of others is less likely to result in abuse of individuals and more likely to represent thinking about a common good. When a researcher acts according to procedures created by patients for their own mutual and collective protection, we do not have paternalism, but self-government.[31] Moreover, control by patients and their families creates the appropriate climate in which to make judgments whether the first criterion—exemplary patient care—is met. This is so because patients' interpretations of the quality of care must be taken into account in such a climate as part of the political process of managing the institution.

Once these two main conditions are met, it would be appropriate and necessary to add conditions regarding suitable procedures for meeting the needs of particular patients. Communities, even if their members are active and even if they benefit most of their members, can function in an oppressive manner. Monitoring the usual four modes of protection described above would still be needed. However, in such settings, criteria for selection of subjects need not rely on the concept of voluntariness, but may focus on the welfare and wishes of the patient.

My suggestion in the Lear story should not be confused with "durable statements of intent"—documents signed in advance of dementia and indicating preferences with regard to care and possible participation in research. Management of care by collective consent of patients resembles durable statements of intent in that it normally involves prior participation by patients in discussions of patient care—perhaps their own. But my approach differs in that it does not require that each patient address each contingency specifically.

Moreover, durable statements of intent only define a document. They do not create a social context capable of preserving the meaning of the document and interpreting it in new circumstances after the patient has declined in mental competence. A durable statement of intent must meet the usual objection: What if the patient seems to change his or her mind and we are not sure of his or her competence? In the appropriate social context, the exact meaning of a prior statement of intent is less important than the overall community understanding of the ongoing character and wishes of the patient. For example, in a close family, the opinion of my sister about what I would now want is probably as valid as my written statement five years ago about what I would now want. The condition of exemplary patient care is meant to be an indication that the social context is suitable for intimate knowledge of the character of patients. The condition of equality of power is offered as a means to that end.

In short, if there is to be invasive research on incompetent SDAT patients, it should only be conducted in institutions that have met basic National Commission requirements (with the exception of voluntariness)[30] and broad and strict criteria with regard to: 1. exemplary care of all patients, and 2. patient participation in institutional management.

General Implications of Nonvoluntary Approaches to SDAT

I am not suggesting that we should rely on informed consent and voluntariness as conditions of research in most cases, and then drop these conditions where we cannot obtain them. I am offering a way of looking at informed consent and voluntariness that emphasizes their social context, not individual patients. That is, it is a good idea to adopt my suggestions with regard to SDAT patients, only it is a good idea to do so generally.

For example, in the case of a fully competent patient expressing an opinion on treatment, we need not think of this opinion as an expression of autonomy or voluntariness in order to take it seriously. Instead, we can take it into account as an important piece of information about the patient's needs and desires. It may even be seen as supremely important, not necessarily because we are interested in the patient's faculty of choice, but because we are interested in the patient's welfare and participation in the management of care.

I have no illusions that my suggestions are easy to implement or that we well understand how to implement them. To the contrary, I believe that through informed consent we ordinarily try hard to find individual solutions to problems that require a much broader social analysis. We turn to the finer individual analyses partly because analysis of broader collective entities is so much more complex. All I have attempted here has been to point in the direction of such an analysis.

Many involved in health care have expressed objections to informed consent procedures. Some of their objections are unconvincing. For example, the claim that "We know what is best for the patient and the patient is better off staying out of it," seldom wins broad support. Yet, some objections are more plausible than others. For instance, there is a lot of information to be considered regarding any study or therapy. It cannot all be given to patients, because patients are limited in their ability to understand medical information, and health professionals are limited in their ability to convey it to patients. Moreover, it may be more important that a certain choice, such as hospice care, be made readily available, than that the ability to consent to it be made available. By pushing the concept of consent to its limits, seeking ethical ways to work with SDAT patients may add strength to those seeking less bureaucratic modes of protecting patients than informed consent procedures.

The Social Context

If King Lear is emblematic of the discussion thus far, Jeremy Bentham is emblematic of what follows. Bentham's body sits in a glass case on display in University College, London. According to some, Bentham was demented in his old age. He has been dead a century and a half, yet he is still given a position of honor and dignity. Whether we have dignity or not is not simply a function of our character; we also bear the mark of our culture and living conditions.[32,33] As our ability to provide for ourselves declines, we show increasingly what society provides for us and less what we provide for ourselves. For Bentham, society is able to provide perpetual public honor; for others, society cannot find clean pajamas. Some societies, for example, support and dignify elders, as shown in the peaceful texture of their faces; other cultures degrade them as shown in their blank and dismal stares.[34,35] Even José Arcadio Buendia, physically deprived and tied to a tree in his family courtyard in all weather, maintained his dignity in spite of advanced dementia.[36] We therefore need to continue to study our perceptions of demented patients in order to sort the social elements of SDAT from the disease elements.

Among the properties of SDAT, mental competence is notably subject to social influence and interpretation. Competence is extremely dependent upon social factors. Like dignity, it is as much a collective and contextual property as an individual attribute. A demented patient trying to eat and a surgeon performing an operation both require social support to function competently. All of us need reminders, calendars, directions, and other social products in order to stay competent. Indeed, as society provides techniques, such as books, to extend our capabilities overall, we may decline in our native abilities, such as memory.[37]

This is another reason that we should look at institutions as a whole in thinking about SDAT research guidelines. The ability of patients to take care of themselves individually depends on the ability of the institution to support them. The criteria of exemplary care and patient participation indicate the ability of institutions to maximize individual patient competence.

Social attitudes toward dementia affect our perceptions of competence. Consider the possible effects of successful research on SDAT. Suppose, for example, that SDAT were easily diagnosable by a simple test. The beginning of Alzheimer's, manifested by memory loss, is virtually indistinguishable from minor, normal memory difficulties.[38] I can just imagine us turning to colleagues who blunder, and suggesting "Really, you should be tested for Alzheimer's." If a test were to return positive, we would have an explanation for these minor failings of competence, and thus an increased likelihood of labeling such mild memory losses as indicative of incompetence. With early detection of Alzheimer's, we can envision such dire consequences as pressure for early retirement, loss of confidence in one's testimony, and the like. Perhaps a test for Alzheimer's would become a condition of tenure. Fantasies aside, a more sensitive test for SDAT could result in more frequent judgments of incompetence and increase the frequency of occasions for showing disrespect rather than increase respect for SDAT patients.

Conclusion

This discussion has consisted mostly in reflection on voluntariness, informed consent, and competence in SDAT. I suggest that the key element of respect for persons in human experimentation—voluntary participation—can be analyzed into a set of related concepts that also express respect for patients. These concepts involve assessing overall patient welfare and collective patient participation. I also suggest that these contextual properties are better understood by appreciating the social factors in perceptions of competence to consent. Incompetence to give consent to participate in experiments arises only partly from disease; it also arises from the inability of institutions to maximize competence and to provide well for patients. Thus, SDAT research in sociology and anthropology should amply supplement research in physiology and neurology. The full extent of human suffering caused by SDAT cannot be understood, nor can a cure be found by addressing the medical issues alone. A disease that so profoundly affects our social relations and the dimensions of which are so profoundly affected by society deserves the broadest possible study.

The debate over informed consent is part of a much older and more extensive debate about the importance of individual human liberty. There are those that see a personal expression of liberty in itself to be vital to human dignity. Others see liberty as less important than a person's ability to obtain what he or she really needs or wants. In this discussion, I do not attempt to resolve this debate. Instead, I attempt to show what it would look like in this case to pursue the view that individual liberty is not as important as what it gets us. In the case of SDAT, this line of thought has some advantages. It makes it possible to include in our sense of what it is to be fully human those whom we regard as unable to make decisions for themselves. We need not ascertain that ineffable borderline between those who are "competent," and those who are "incompetent." Instead, we substitute for it a concept of human community that is assiduous in protecting the welfare of all its members through the active participation of all.

Acknowledgments

I appreciate the careful comments and support of Christine K. Cassel MD, and Robin Linden.

Notes and References

[1]A. R. Jonsen and M. Yesley, "Rhetoric and Research Ethics: An Answer to Annas," *Medicolegal News* 8(6):1980, 8–13.

[2]R. Adolfsson, "Prevalence of Dementia Disorders in Institutionalized Swedish Old People: The Workload Imposed by Caring for these Patients," *Acta Psychiatrica Scandinavia* 63, 1981, 225–244.

[3]The Alzheimer's Disease Center, Albert Einstein College of Medicine and National Institute on Aging, National Institutes on Health, *Alzheimer's Disease: Q & A* (Washington, DC: US Government Printing Office, 1980).

[4]R. N. Butler, "Aging: Research Leads and Needs," *Forum on Medicine*, November 1979, 716–725.

[5]S. G. Haynes et al., eds. *Proceedings of the Second Conference on the Epidemiology of Aging: March 28–29, 1977,* National Institutes of Health (Washington, DC: US Government Printing Office, 1980).

[6]J. W. Jacobs et al, "Screening for Organic Mental Syndromes in the Medically Ill," *Annals of Internal Medicine* 86, 1977, 40–46.

[7]R. Katzman, "The Prevalence and Malignancy of Alzheimer's Disease: A Major Killer," *Arch Neurol* 33, 1976, 217–218.

[8]J. E. Bernstein and F. K. Nelson, "Medical Experimentation in the Elderly," *Journal of American Geriatric Society* 23(7), 1975, 327–329.

[9]R. N. Butler, "Protection of Elderly Research Subjects," *Clinical Research* 28(1), 1980, 3–5.

[10]S. de Beauvoir, *The Coming of Age,* trans. P. O'Brian (New York: Warner Books, 1973).

[11]E. E. Lau and J. I. Kosberg, "Abuse of the Elderly by Informal Care Providers," *Aging,* September–October, 1979, 10–15.

[12]A. Norberg, B. Norberg, and G. Bexell, "Ethical Problems in Feeding Patients with Advanced Dementia, *British Medical Journal* 281, 1980, 847–848.

[13]M. Sim, E. Turner, and W. T. Smith, "Cerebral Biopsy in the Investigation of Presenile Dementia: I. Clinical Aspects," *British Journal of Psychiatry* 112, 1966, 119–125.

[14]W. T. Smith, E. Turner, and M. Sim, "Cerebral Biopsy in the Investigation of Presenile Dementia: II. Pathological Aspects," *British Journal of Psychiatry* 112, 1966, 127–133.

[15]L. R. Caplan, M. S. Hirsch, and M. N. Swartz, "Brain Biopsy in Herpes Simplex Encephalitis," *New England Journal of Medicine* 303(12), 1980, 700–701.

[16]P. E. Etienne et al., "Adverse Effects of Medical and Psychiatric Workup in Six Demented Geriatric Patients," *American Journal of Psychiatry* 138(4), 1980, 520–521.

[17]J. J. McCartney, "Encephalitis and Ara-A: An Ethical Case Study," *The Hastings Center Report* 8(6), 1978, 5–7.

[18]P. Ramsey, *The Patient as Person* (New Haven: Yale University Press, 1970).

[19]E. R. Gonzalez, "Psychosurgery: Waiting to Make a Comeback?" *JAMA* 244(20) 1980, 2245–2251.

[20]W. W. Lowrance *Of Acceptable Risk: Science and the Determination of Safety* (Los Altos, California: William Kaufmann, 1976).

[21]National Institute of Neurological and Communicative Disorders and Stroke (NINCDS), *The Dementias: Hope through Research* (Washington, DC: US Government Printing Office, 1981).

[22]National Institute on Aging, *Recent Developments in Clinical and Research Geriatric Medicine: The NIA Role* (Washington, DC: US Government Printing Office, 1980).

[23]Task Force sponsored by the National Institute on Aging, "Senility Reconsidered: Treatment Possibilities for Mental Impairment in the Elderly," *JAMA* 244, 1980, 259–263.

[24]E. Baumgarten, "The concept of 'competence' in Medical Ethics," *Journal of Medical Ethics* 6, 1980, 180–184.

[25]P. A. Freund, ed., *Experimentation with Human Subjects* (New York: George Braziller, 1969, 1970.

[26]A. Meisel, L. H. Roth, and C. W. Lidz, "Toward a Model of the Legal Doctrine of Informed Consent," *American Journal of Psychiatry* 134(3), 1977, 285–289.

[27]The National Commission for the Protection of Human Subjects of Biomedical and Behavioral Research, *Appendix to Report and Recommendations: Psychosurgery* (Washington, DC: US Government Printing Office, 1977).

[28]*Report and Recommendations: Psychosurgery* (Washington, DC: US Government Printing Office, 1977).

[29]*Report and Recommendations: Research Involving Children* (Washington, DC: US Government Printing Office, 1977).

[30]*Report and Recommendations: Research Involving Those Institutionalized as Mentally Infirm* (Washington, DC: US Government Printing Office, 1978).

[31]G. Dworkin, "Paternalism," in S. Gorovitz et al. *Moral Problems in Medicine* (Englewood Cliffs, NJ: Prentice-Hall, 1976).

[32]J. A. Smithers, "Institutional Dimensions of Senility," *Urban Life* 6(3), 1977, 251–276.

[33]W. F. Thomas, "Aging Disorders Reflect Social Deprivation," Unpublished manuscript, Department of Behavioral Science, Tarrant County Junior College, South Campus, Fort Worth, Texas.

[34]J. S. Kayser-Jones, *Old, Alone, and Neglected: Care of the Aged in Scotland and the United States* (Berkeley: University of California Press, 1981).

[35]P. A. Stathopoulos "Nursing Homes and Consumer Advocacy," *Generations* 4(1), 1980, 28–29, 64.

[36]G. Marquez, *One Hundred Years of Solitude* (New York: Avon Books, 1970).

[37]H. Braverman, *Labor and Monopoly Capital: The Degradation of Work in the Twentieth Century* (New York: Monthly Review Press, 1974).

[38]E. H. Liston "The Clinical Phenomenology of Presenile Dementia: A Critical Review of the Literature," *Journal of Nervous and Mental Disease* 167(6), 1979, 329–336.

Technical Aspects of Obtaining Informed Consent from Persons with Senile Dementia of the Alzheimer's Type

Richard M. Ratzan

Although several authors have recently championed the cause of informed consent in clinical geriatrics,[1–4] few have dealt with specific, nuts-and-bolts issues involving technique. And yet, as any mouse who has tried to bell the cat can tell you, once the idea mice have had their say, technique is everything. In this paper I shall treat such technical issues as presentation of content, informed consent forms, and perceptual adjuncts to facilitate the communication of the proposed experiment or therapy.

Clarity

In communicating the content of the proposed experiment, the researcher should aim for clarity. Although clarity of presentation is no longer a novel element of informed consent, it is, however, still frequently overlooked by professionals immersed in a technical vocabulary as arcane to patients and subjects as the astronauts' space vocabulary is to non-astronauts. Unfortunately, the researcher does not usually employ an expert commentator, as NBC does for its coverage of our space shots, to translate the jargon emanating from mission control.

Clarity means no jargon. Especially no medical jargon. Mary Wolanin aptly demonstrates the disservice researchers perform when they communicate in a fashion other than that which William Carlos Williams, himself a physician with forty years' service as a practitioner, called "speaking straight ahead"[5]: "You would say 'poor prognosis,' " Ms Wolanin writes, "but for those of us who are doing it, just say 'going downhill.' We know what that means."[6] John McKinlay has demonstrated that some physicians commit the unforgivable error (in any sort of communication) of using language that they themselves, at the time they were presenting it to them, doubted their clinic patients understood.[7]

Risk

I have written more fully about risk in clinical geriatric research elsewhere. In this section I would like to highlight the salient features of this subject and to elaborate a few particulars. The general topic of risk has been amply investigated by William Lowrance.[8] He divides his analysis of risk into risk assessment: a quasi-objective, statistical quantification (the world of facts and numbers) and risk acceptability: a subjective qualification based on emotions and judgment (the world of values). Bertram Dinman has applied this type of two-step analysis to risk vis-à-vis the maintenance of health and certain aspects of occupational medicine.[9] He suggests that "risks that are undertaken voluntarily or are imposed by the simple fact of living can be recast in commonly acceptable terms encountered in everyday life."[10] He illustrates this recasting by contrasting the risk of death from such voluntary activities as horse racing (1:740), automobile driving (1:5900), and early legal abortion (1:50,000) to death from such involuntary phenomena as leukemia (1:12,500), falling aircraft (1:10 million, US), and floods (1:455,000).

This translation of the comparative taxonomy of risks into practical terms has much to recommend it. First, it avoids the semantic opacity and ambiguity of adjectival phrases like "greater than minimal risk," a phrase that even logical positivists might shudder to define, much less elderly SDAT subjects. Secondly, it offers an almost tangible means for the patient or subject to use as a basis of comparison. As Loretta Kopelman so aptly illustrates, a good working semantic definition of minimal risk as originally defined by DHEW, i.e., "the probability and magnitude of harm that is normally encountered in the daily lives of healthy individuals," is far from a consensual reality for all researchers and all subjects who must use such a definition.[11]

I recommend that a list of involuntary risks like Dinman's be compiled from the sundry sources scattered throughout the geriatric and US Census literatures.

Voluntary risks remain the same for elderly persons in one respect, yet different in another. They remain the same insofar as the elderly subject is the average US or UK citizen used in the compilation of statistics cited above. Voluntary risks differ, however, insofar as many elderly people change their daily activities. The voluntary risk of dying in an automobile accident approaches zero for the elderly person who has chosen to eliminate this risk or whose ophthalmologist strongly discourages it after bilateral cataract surgery with less than optimal results. Stated differently, "Those who live long enough to become elderly often do so by decreasing the number of risks in their lives, for they know they have little or no control over the magnitude of these risks. . . . A cold may lead to pneumonia," a splinter to death in the hospital.[12] In other words, elderly "conservative" risk-takers have lowered their levels of risk acceptability after intuitive, often life-long assessments of such risks. "Minimal risk" as defined by daily life, consequently, becomes a more stringent standard for such elderly SDAT subjects and should not be inferred by the clinical researcher to equal what a 40 year old, or elderly "liberal" risk-taker, considers "minimal risk."

There are, however, problems with recommending that a researcher compare the risk of a certain experimental complication to the risk of, say, death from leukemia or death from a fall. First of all, such quantification is often not known for the former. Second, it is often not immediately accessible for the latter. But whenever possible, an attempt should be made, in honest terms, of equating a low-probability complication, like death from contrast radiography, to a risk in daily living. If it is not known, then the researcher ought to say so, embracing Dinman's position that a "full disclosure includes an exposition of what is *not* known or is unclear."[9] The recent report by the President's Commission for the Study of Ethical Problems in Medicine and Biomedical and Behavioral Research emphasizes this obligation and summarizes a study of physician and patient attitudes towards the discussion of uncertainty in health care.[13]

The clinical researcher must respect the fact that in any experiment not aimed at therapy, any risk is additional risk, an important reminder Loretta Kopelman has elucidated.[11] The ethically correct approach, therefore, should be what Cyril Comar calls a "pragmatic de minimis approach."[14] Such an approach eliminates any easily avoidable risk or one without benefit; eliminates any large risk without "clearly overriding benefits"; "ignores for the time being any small risk"; and actively studies risks in between small and large risks. However, since Comar defines "small" as "about 1 in 100,000 per year or less," I suggest we eliminate the guideline to ignore "small" risks when considering research not aimed at therapy, especially when the subject is mentally competent. It should be the subject's option to ignore or not to ignore odds of any magnitude.

For research aimed at therapy, perhaps Comar is right. I disagree with his reasoning, however. The statement that the "hard fact is that attempts to eliminate risks for the unfortunate few tend to markedly increase them for the rest of a large population" smacks of paternalistic intervention—an attitude unfair to those SDAT subjects with no one to defend them but IRBs in the instance of minimal risk,[11] or only the potentially inadequate and variably motivated proxy consent of a relative or significant other.[12]

The opposite attitude, i.e., paternalistic protection from risk, should not be inferred. The mentally competent subject who is willing and desirous of engaging in risky research has a right to do so, and should not be forbidden this privilege simply because he or she is old, infirm, and in an institution. John Crowe Ransom understood this dangerous assumption very well and expressed it eloquently in his poem, "Old Man Playing with Children."[15] Listen to the grandfather mentally addressing his overprotective son:

> Do not offer your reclining chair and slippers
> With tedious old women talking in wrappers.
> This life is not good but in danger and in joy.

The Informed Consent Form: To Be Or Not To Be

The last ethical locus for potential researcher error that I shall treat in this paper is the informed consent form. Although the clinician or clinical researcher often fondles this signed piece of paper like a child fondling his magically protective blanket, several qualified observers of informed consent have recently brought to the attention of these magical thinkers the fallacy of such a belief (summarized by Herbert 16, see ref.). As a matter of fact, the magic blanket may, as Leslie Miller has pointed out,[17] turn out to be a Damoclean sword held over the physician's head as evidence of fraud. I believe that there is a need for the informed consent form in clinical geriatric research and shall present my argument for it in terms of the increased likelihood of a truly informed consent. Such a likelihood is increased, however, only if the informed consent form is easily understandable. This likelihood is further augmented if the form also includes a second part to test comprehension. Furthermore, such an improved informed consent form can become the most important part of a larger educational process aimed at the special perceptual and cognitive attributes of elderly research subjects.

In the last few years there has been a flurry of interest in the general process, and the prose style in particular, of informed consent forms.[18–20] The answer to "Why can't Johnny read the consent forms?" is now clear: Johnny can't read the consent forms because they weren't written for or by people like Johnny. The old truism in teaching, i.e., if a teacher wants to find out how well he has constructed a test he has only to give it, has finally come home in clinical research. The recent literature relating subjects' failures on post-informed consent form tests, briefly summarized by Rennie,[21] makes it evident that, regardless of which party is more responsible, teacher or student, clinical researchers must address the problem of how best to impart the information part of the informed consent process with or without using informed consent forms. The existence of the problem has been discovered. The solution hasn't.

Readability

"Readability" refers to the ease with which a person reads and comprehends a text. In this discussion, the text is an informed consent form. Some suggestions for constructing more easily readable informed consent forms are as follows:

1. The informed consent forms should be short. Epstein and Lasagna demonstrated that, in a mock experimental situa-

tion studying the informed consent process, "comprehension and consent to volunteer were inversely related to length of forms."[22] Virchow realized this important dictum in prose style long before Strunk and White's *The Elements of Style* advised writers, among other things, to "omit needless words" and not to "explain too much." Virchow wrote, "Brevity in writing is the best insurance for its perusal." Or, to paraphrase Voltaire, "If a clinical researcher had more time (and he should make it), he'd write a shorter informed consent form."

2. The print of the form should be in large type in consideration of the high incidence of presbyopia in any geriatric research population. Harvey Taub, in his excellent study of the relationships between memory and informed consent in elderly subjects, used the Orator 10 type.[23]
3. Write the forms with a style and diction that are easily readable by the subjects for whom they were intended. Multiple studies have demonstrated an inappropriate level of complexity (diction, sentence structure, word length, sentence length) of informed consent forms encountered across the country.[18,20,24] T. M. Grundner has published a useful summary of the two complementary formulas for determining readability: the Flesch Readability Formula and the Fry Readability Scale.[25] Informed consent forms should not be, as they have been demonstrated to be, written for IRBs and other researchers. The purpose of informed consent forms is to inform. Uninterpreted phrases like "palliative radiotherapy" (to indicate an "alternative treatment") have no place on an informed consent form. "Avoid fancy words" advise Strunk and White. This advice is as good for Cornell undergraduates writing essays as it is for clinical researchers composing informed consent forms.

If, as Barrett and Wright have proposed, aged subjects show "higher recall than younger adults when subjects [process] words more familiar to older adults," then perhaps we should consider using words like "poultice" instead of "dressing" on informed consent forms for older adults.[26] This suggestion may be particularly helpful for SDAT patients who live in a world of uncertain vocabularies, trying their hardest to communicate with dimly perceived shadows for words. As Sally Gadow has suggested, "the facile assumption of incompetence in many elderly patients often reflects only the lack of professional energy, interest, or imagination needed to communicate with persons of different mentation than 'normal.' "[27] It does require energy and work to communicate with a cognitively impaired SDAT research subject. However, unless the clinical researcher is willing to expend that energy, he is performing an experiment on, not with, that subject.

4. As Gary Morrow has recommended, the clinical researcher should have his informed consent forms "read and critiqued by patients, not fellow professionals" for possible areas of confusion. (However, the researcher would soon discover that Morrow's immediately preceding suggestion, i.e., "Write as you talk," would be counterproductive for most clinical researchers for the reasons discussed above. See refs.[6,7]) This dry run of the informed consent form finds its equivalent, in modern day pedagogy, in trial test questions preceding the final administration of any large-scale test by a major educational testing service. If we care enough to give fair tests to high school and college students when only grades are at stake, shouldn't we care enough to give fair tests to their grandparents when their dignity and comfort and health are at stake?

The Two-Part Consent Form

In 1974 Robert Miller and Henry Willner proposed a radically new consent form that was as ingenious as it was obvious.[28] Since traditional informed consent forms condense two processes—the explanation of information and the subject's signed statement that he understands it and consents to it—Miller and Willner proposed a separation of these two processes into two steps to ensure the subject's comprehension of the information. They therefore proposed that the second part, a questionnaire assessing the subject's actual understanding, include questions concerning "benefits; departures from ordinary medical practice; risks; inconveniences and tasks; purposes and rights." If the subject "passed" the test to the satisfaction of the investigator or another less-biased grader, then the investigator would leave a copy of the "test" with the subject. If the subject failed, Miller and Willner proposed the following options: the subject could retake the test after studying the informed consent form; he could retake the test again after an oral explanation; or he could be disqualified after repeated failures. Miller and Willner offered as benefits, in addition to a more accurate assessment of the subject's actual understanding, increased time for the subject to make a decision about consent and the researcher's assessment of the efficacy of his informed consent form in conveying information.

I have devoted perhaps more space than in necessary for such a well-known gem in the informed consent literature because it is still an unused gem. I have yet to see, nine years after its proposal, a two-part consent form. It is, and has been, more admired in the breach than in practice. Perhaps this is true because of the apparently excessive expenditure of time, energy, and cost in implementing such a proposal despite its evident need.

Several studies have amply demonstrated the need for a two-step approach. Roger Williams et al. instituted such a two part process, reporting their findings in 1977.[29] Their data are nothing less than shocking. In an experiment involving the inoculation of prisoners with malaria, only six of twenty "control" volunteers appreciated their risk of contracting potentially fatal hepatitis as a result of the study, whereas fourteen of the second group of twenty volunteers—volunteers who had had the experiment explained to them by a physician—appreciated the risk. Gary Morrow et al. showed an improvement in cancer patients' understanding of appropriate alternatives, procedures, and their diagnoses when tested one to three days after looking over their informed consent forms as compared to those patients tested immediately.[30] Likewise, Amelia Schultz et al.,[31] George Robinson and Avraham Merav,[32] and Barrie Cassileth et al.[19] have all reported dismal scores for patients taking tests variable lengths of time following the explanation of the study of their signing the informed consent forms.

Data for elderly subjects in these and other similar studies are meager and tenuous, but suggestive nonetheless. As mentioned above, Morrow et al. found a tendency for a greater degree of informed consent amongst the younger patients than amongst the older ones.[30] Barrie Cassileth et al., in their study of two hundred cancer patients and their consent for cancer therapy, discovered that patients over the age of 65, if they read them at all, read their informed consent forms in the most careless fashion of all groups.[19] Since education, the great confounder of cognitively charged covariables, was also significantly associated with care of reading, this association of age must remain only a suggestive one.

Possibly corroborating this finding of an apathetic reading of informed consent forms by elderly subjects is the paper by James Sands and John Parker, who studied perceived stressfulness of various life events.[33] An unexpected discovery was the "interesting finding . . . that significantly fewer elderly persons completed the rating of all items on the questionnaire. This was true in spite of the fact that of the three groups, the elderly men received the most instructional assistance." In this last study, as Jack Botwinick's studies would predict, the elderly subjects tended to opt not to opt when given that option[34] (see also ref. 4). The implications are evident and offer a strong argument for serial testing of informed consent in SDAT subjects prior to the initiation of the research. In the school of informed consent, "no pass" must mean "no research."

Alan Meisel has posted his disagreement with this last principle, arguing that "to require that the patient understand the disclosed information—whatever that may mean—is at odds with patient autonomy, because patients who are deemed not to understand will be denied the right to make their own decisions."[35] This admirable antipaternalistic objection, however, must be carefully applied to one, and not the other, of the two possible scenarios for an elderly subject who has failed his informed consent test.

The first scenario is the elderly subject who would like to pass but is cautious and suffering from lack of self-confidence; who doesn't know what "alopecia" means but is afraid to ask; who doesn't remember what "paralyzed" means but is afraid to ask; who fails because he feels he is expected to fail; and so forth. This subject should not be allowed to go quietly into that good experiment. He needs a skilled interviewer who will assiduously plumb the depths and breadths of his "ignorance" in order to ascertain why he fails. Researchers who use subjects who have involuntarily failed informed consent tests are failing their own medical ethics tests. I do not think that Meisel is referring to this subject, but his stance could include such a subject, as it now reads, without further clarification.

The second scenario is the subject who fails and doesn't mind failing informed consent tests, or doesn't want to pass. This is the patient who tells the doctor, "You decide. You're the doctor. Whatever you say is okay with me." In agreement with Meisel, and in opposition to Buchanan,[36] I recognize the existence of such doctor–patient relationships. Even the original antipaternalist, John Stuart Mill, allowed himself the indulgence of a non sequitur when he stated that "The principle of freedom cannot require that he should be free not to be free. It is not freedom to be allowed to alienate his freedom."[37] On the contrary, that is the exact epitome of freedom, and is the reason why at one and the same time we must allow the elderly research subject to enact scenario number two for one experiment, no matter how dangerous it is, and enact scenario number one for another, even if he ultimately refuses to volunteer for an experiment that he understands perfectly well to entail (were it possible) zero risk. Karen Lebacqz has also drawn this two-scenario distinction in her recommendations concerning nondisclosure of information in neuropsychopharmacological research, much of which is done using elderly subjects.[38]

The final bit of evidence arguing for the need for a two-step informed consent process in the elderly research subject comes from Harvey Taub's exemplary study of the effects of memory, age, and vocabulary levels on informed consent.[3] This study compared the scores of 34 young, non-institutionalized women and 56 non-institutionalized elderly women on a test measuring their memories for the information they had read on informed consent forms two to three weeks earlier. The experiment within the experiment concerned information delivery. Some of the results may be summarized as follows: there was a direct effect of vocabulary on performance in the elderly subjects; performances of old and young women of identical vocabulary levels were similar; and all participants at all vocabulary levels had relatively poor cumulative scores. For example, 47 to 69% of the participants at the highest vocabulary levels only answered three of the five questions correctly. There was no difference between the subgroups who read informed consent forms rated as "fairly difficult" by the Flesch readability score and those who read the "fairly easy" forms.

The two part consent form, as Taub suggests, may be useful for elderly research subjects. I believe the successful circumvention of the "ageism" objection, i.e., that restricting the use of two-part informed consent forms to elderly research subjects alone is paternalistic, is an easy one. Making two-part informed consent forms standard fare for all research subjects—and the studies above all support such an age-independent need—

would protect everyone from the vagaries of a distracted, emotionally taxed, or individually poor memory.

Making all informed consent forms two-part forms could yield diagnostic and therapeutic advantages. First, they can objectively assess, i.e., diagnose, a subject's actual understanding of the information necessary for a valid consent. Insofar as the standard consent form is intended to do this, it hardly seems paternalistic to ensure that it in fact does do this, and not just in intention. The road to invalid consent is paved with unassessed intentions. Alan Meisel et al. have underlined the real discrepancy that can exist between intentions and results, reminding us that "the act of informing someone does not assure that one will understand the information that has been imparted."[39] This discrepancy hardly seems academic in a group of subjects who, so far at least, have evidenced a less than encouraging interest in reading the information present on informed consent forms.[19]

A second, related benefit, potentially both diagnostic and therapeutic, is the researcher's ability to use the second part, i.e., the test, as a tool both to measure ongoing education of the subject until actual understanding is reached and to direct such education at the specific questions being answered incorrectly.

A third, diagnostic benefit allows the clinical researcher to assess his own actual understanding of how best to compose informed consent forms. The two part consent form is, therefore, a risk-free insurance policy, a pretest as it were, allowing both researcher and subject the opportunity to perform a mental experiment before the real one has indicated insufficient understanding at possibly excessive expense.

A fourth advantage is the built-in delay. Such a delay between the provision of information and the subject's response to it allows the two-part consent form to be a two-way street, i.e., communication. To allow sufficient time for the decision-making process to be conducted at a rate comfortable for an elderly subject is crucial. Jack Botwinick has reviewed the literature on the interaction between total time involved in leaning and age, and suggests that "the speed with which an older person can respond limits his ability to demonstrate what he has learned, but in itself is not regarded as a cognitive factor."[34] Botwinick's statement that "it takes longer for older people to learn material than it does for younger people" makes a strong argument for the beneficial effects of the delay inherent in the two-part informed consent form.

Learning and rate of learning, however, are but one pair of variables potentially improved as a result of delay. Cognitive disadvantages such as poor memories and slower memories[40]; attitudinal constraints such as cautiousness and the hopelessness of institutionalization; peer pressures such as the ones suggested in my paper[12] and Taub's[3]; the fluctuating mental status of SDAT subjects because of their primary disease, or psychiatric disease, or drugs—or all three; the desire to discuss the proposed experiment with family members and/or to ask the researcher questions—all militate for a delay between "informed" and "consent."

A final benefit is the possible practice-effect that such a test may have, especially if it is necessary to repeat it several times. Although the data concerning the effect of practice on cognitive performance in the elderly is conflicting,[41] studies by Harvey Taub[42] and David Hultsch[43] have demonstrated a positive effect of practice on memory and learning in the elderly. As Nancy Denney suggests, "practice may be more beneficial for some types of abilities than for others."[41]

In the last analysis, however, Victor Herbert correctly reduces the whole issue of "informed consent forms or no informed consent forms" to its proper dimensions: "The fact that the patient gave an informed consent usually will not prevent him from suing; a warm relationship with a competent and caring physician usually will."[16]

Procedural Adjuncts to Obtaining an Informed Consent

This last section suggests some ways in which a clinical researcher who is truly desirous of obtaining an informed consent may revise the traditional one-on-one, single interview process into the format most suitable for a particular subject.

1. Leave a copy of the informed consent form with the subject.

Gary Morrow et al. found this technique very useful in increasing the amount of information that the "take-home" patients had when compared to control patients.[30] This "delayed group" had one to three days to review the information. Miller and Willner also recommend this technique for two reasons. First, it would allow the subject to re-examine his involvement in the experiment in a knowledgable way, with or without help from friends and relatives. Second, it can "help the subject detect differences between his actual experience in the experiment and what he expected. He can then bring any discrepancies to the attention of the investigator."[28] The well-known benefit of giving the student homework before the test is, unfortunately, still news to some clinical researchers.

2. Encourage significant others to be present during the interviews.

Robert Moore suggests that, in addition to having close family members present, the researcher should encourage them to sign the informed consent forms as witnesses.[44] Hugh Butt discovered that one benefit of his technique of leaving a taped recording of the interview with the patient was that three quarters of the 45 patients polled responded that they "found it helpful to have their spouse [sic] or relatives listen to the tape."[45] Note that the average age was 55.

As Moore puts it, "as a practical matter, they [close family members] often assist the patient in his decision about the procedure." The elderly subject often does want family help and this desire ought not only to be respected, it ought also to be encouraged and enlisted when the subject does desire it. To force freedom of choice on an elderly subject who does not want freedom of choice is coercive paternalism and deprives him of his freedom to choose dependence.

3. Perceptual adjuncts can be helpful.

Hugh Butt found tape recording very useful in increasing the amount of information that the subjects had about the experiment.[45] Galen Barbour used videotapes of the original explanation of self-care dialysis programs to augment patient understanding and to educate patients in areas of continued ignorance.[46] In a study of particular significance to the issue of elderly subjects' ability to learn new information, William Woodward et al. used slides as a supplementary technique to explain cholera vaccines to volunteers.[47] The results indicated a significant increase in the information obtained by the volunteers seeing the slide show. This technique may prove especially helpful to elderly research subjects since many students of cognition in the elderly seem to agree on two points. First, that elderly subjects do not spontaneously use strategies to learn new material and/or to solve problems; second, that when instructed to use mediational techniques, especially visual imagery, elderly subjects often show dramatic improvement in their cognitive performances.[34] Since verbal mediational techniques have also proved helpful, tape recordings may likewise augment elderly subjects' learning of the information about proposed experiments. Videotapes, i.e., the combination of verbal and visual mnemonic techniques, might even prove synergistically beneficial.

4. Communication means two signatures on the informed consent form.

The word "communicate" is derived from the Latin verb "communicare," or "to make common; put into a common stock; share." Caesar wrote of "communicating," i.e., sharing, glory with his legions. Cicero used it in referring to the communication of, i.e., the sharing of, Roman citizenship with others. Such a prefixal use of the prepositional "com" denotes doing that particular verb "with" someone. For "communication," it means doing a munus, i.e., a "service, function, or duty" with someone else.

One can only "communicate," i.e., "share stock with," if there is another person to share it with. This sharing of one's "munus" means, for both researcher and subject, the sharing of a duty to cooperate with each other and a duty to respect one another. The clinical researcher, insofar as he is able, attempts to serve that subject and that subject's disease and, indirectly, other subjects, especially subjects with that specific disease. The subject serves the researcher and the research process and indirectly, other subjects, especially with the same disease.

The informed consent form, therefore, ought to epitomize what James Vaccarino has called "the essence of rapport" and should include both the investigator's and the subject's signatures, signed in each other's presence.[48] This co-signing is a leveling principle, a joint affirmation that clinical research is an ethically invested service of mutuality.

Acknowledgment

I would like to dedicate this paper to the memory of my Uncle Lawrence A. Engel who, at the age of 91, realized his desire last year to "die young at a very old age."

References

[1]W. T. Reich, "Ethical issues related to research involving elderly subjects," *Gerontologist* 18, 1978, 326–327.

[2]J. L. Makarushka, and R. D. McDonald "Informed consent, research, and geriatric patients: the responsibility of institutional review committees," *Gerontologist* 19, 1979, 61–66.

[3]H. A. Taub, "Informed consent, memory and age," *Gerontologist* 20, 1980, 686–690.

[4]R. M. Ratzan, "Cautiousness, risk, and informed consent in clinical geriatrics," *Clinical Research* 30, 1982, 345–353.

[5]G. Plimpton, ed., *Writers at Work*, 3rd series, New York: Penguin, 1977, 3–30.

[6]M. O. Wolanin, "Communication with the aged," in H. B. Haley, and P. A. Keenan, eds., *Health Care of the Aged*, Charlottesville: University Press of Virginia, 1981, 100–109.

[7]J. B. McKinlay "Who is really ignorant—physician or patient?" *Journal of Health and Social Behavior* 16, 1975, 3–11.

[8]W. W. Lowrance, *Of Acceptable Risk*, Los Altos, California: Wm Kaufmann, 1976.

[9]B. D. Dinman, "Occupational health and the reality of risk—an eternal dilemma of tragic choices," *Journal of Occupational Medicine* 22, 1980, 153–157.

[10]B. Dinman, "The reality and acceptance of risk," *Journal of the American Medical Association* 244, 1980, 1226–1228.

[11]L. Kopelman, "Estimating risk in human research," *Clinical Research* 29, 1981, 1–8.

[12]R. M. Ratzan " 'Being old makes you different': the ethics of research with elderly subjects," *Hastings Center Report* 10, 1980, 32–42.

[13]"Making health care decisions: the ethical and legal implications of informed consent in the patient-practitioner relationship." Volume one:report. President's Commission for the study of ethical problems in medicine and biomedical and behavioral research. October 1982. (For sale by the Superintendent of Documents, US Government Printing Office, Washington, DC 20402.)

[14]C. L. Comar "Risk: a pragmatic de minimis approach," *Science* 203, 1979, 319.

[15]J. C. Ransom, "Old man playing with children," in O. Williams, and E. Honig, eds., *The Mentor Book of Major American Poets*, New York: Mentor, 1962, 397.

[16]V. Herbert, "Informed consent—a legal evaluation," *Cancer* 46, 1980, 1042–1044.

[17]L. J. Miller, "Informed consent: III," *Journal of the American Medical Association* 244, 1980, 2556–2558.

[18]G. R. Morrow, "How readable are subject consent forms?" *Journal of the American Medical Association* 244, 1980, 56–58.

[19]B. R. Cassileth, R. V. Zupkis, K. Sutton-Smith, and V. March, "Informed consent—why are its goals imperfectly realized?" *New England Journal of Medicine* 302, 1980, 896–900.

[20]T. M. Grundner, "On the readability of surgical consent forms," *New England Journal of Medicine* 302, 1980, 900–902.

[21]D. Rennie, "Informed consent by "well-nigh abject" adults," *New England Journal of Medicine* 302, 1980, 917–918.

[22]L. C. Epstein, and L. Lasagna, "Obtaining informed consent," *Archives of Internal Medicine* 123, 1969, 682–688.

[23]H. A. Taub, "Informed consent, memory and age." *Gerontologist* 20, 1980, 686–690.

[24]B. H. Gray, R. A. Cooke, and A. S. Tannenbaum, "Research involving human subjects," *Science* 201, 1978, 1094–1101.

[25]T. M. Grundner, "Two formulas for determining the readability of subject consent forms," *American Psychologist* 33, 1978, 773–775.

[26]T. R. Barrett, and M. Wright, "Age-related facilitation in recall following semantic processing," *Journal of Gerontology* 36, 1981, 194–199.

[27]S. Gadow, "Medicine, ethics, and the elderly," *Gerontologist* 20, 1980, 680–685.

[28]R. Miller, and H. S. Willner, "The two-part consent form" *New England Journal of Medicine* 290, 1974, 964–966.

[29]R. L. Williams, K. H. Reickmann, G. M. Trenholme, et al., "The use of a test to determine that consent is informed," *Military Medicine* 142, 1977, 542–545.

[30]G. Morrow, J. Gootnick, and A. Schmale, "A simple technique for increasing cancer patients' knowledge of informed consent to treatment," *Cancer* 42, 1978, 793–799.

[31]A. L. Schultz, G. P. Pardee, and J. W. Ensinck, "Are research subjects really informed?" *Western Journal of Medicine* 123, 1975, 76–80.

[32]G. Robinson, and A. Merav, "Informed consent: recall by patients tested postoperatively," *Annals of Thoracic Surgery* 22, 1976, 209–212.

[33]J. D. Sands, and J. Parker, A cross-sectional study of the perceived stressfulness of several life events, *International Journal of Aging & Human Development* 10, 1979–1980, 335–341.

[34]J. Botwinick, *Aging and Behavior*, New York: Springer, 1978.

[35]A. Meisel, "Consent forms," *Journal of the American Medical Association* 245, 1981, 921–922.

[36]A. Buchanan, "Medical paternalism," *Philosophy & Public Affairs* 7, 1978, 369–390.

[37]J. S. Mill, *On Liberty,* Chicago: Gateway Editions Inc (distributed by Henry Regnery Co):131. (Originally published in 1859.)

[38]K. A. Lebacqz, "Ethical issues in psychopharmacologic research," in D. M. Gallant, and R. Force, eds. *Legal and Ethical Issues in Human Research and Treatment,* New York: Spectrum Publications (distributed by Halsted Press Division of John Wiley & Sons, Inc.) 1978, 113–142.

[39]A. Meisel, L. H. Roth, and C. W. Lidz, "Toward a model of the legal doctrine of informed consent," *American Journal of Psychiatry* 134, 1977, 285–289.

[40]T. R. Anders, J. L. Fozard, "Effects of age upon retrieval from primary and secondary memory," *Developmental Psychology* 9, 1973, 411–415.

[41]N. W. Denney, "Adult cognitive development," in D. S. Beasley, and G. A. Davis, eds., *Aging: Communication Processes and Disorders,* New York: Grune & Stratton, 1981, 123–137.

[42]H. A. Taub "Memory span, practice, and aging," *Journal of Gerontology* 28, 1973, 335–338.

[43]D. F. Hultsch, "Learning to learn in adulthood," *Journal of Gerontology* 29, 1974, 302–308.

[44]R. M. Moore, Jr., "Consent forms—how, or whether, they should be used," *Mayo Clinic Proceedings* 53, 1978, 393–396.

[45]H. R. Butt, "A method for better physician–patient communication," *Annals of Internal Medicine* 86, 1977, 478–480.

[46]G. L. Barbour, and M. J. Blumenkrantz, "Videotape aids informed consent decision," *Journal of the American Medical Association* 240, 1978, 2741–2742.

[47]W. E. Woodward, "Informed consent of volunteers: a direct measurement of comprehension and retention of information," *Clinical Research* 27, 1979, 248–252.

[48]J. M. Vaccarino, "Consent, informed consent and the consent form," *New England Journal of Medicine* 298, 1978, 455.

Part 4

Institutional Issues

The Need for Alternatives to Informed Consent by Older Patients

Psychological and Physical Aspects of the Institutionalized Elderly

Pamela B. Hoffman and Leslie S. Libow

Introduction

Issues revolving around the process of informed consent for research on elderly subjects are compounded for the physician dealing with institutionalized patients.[1–10] The effects of institutionalization on a subject's ability to give consent must be considered: the population of a nursing facility not only constitutes a representative community of subjects for research, but it also includes a large number of dementia patients, particularly those with moderate and severe dementias. Hence, the nursing home is a crucial element in geriatric research. Since research on senile dementia of the Alzheimer's type (SDAT) must include the nursing home population, it is necessary to review the characteristics of the institutionalized elderly that make them psychologically and physically vulnerable to research.

The Institutional Environment

The nursing home represents an environment quite distinct from the usual research setting found in hospitals or the community. Nursing homes in the United States are predominantly proprietary facilities and do not have a tradition of involvement with research. The majority of residents are women, often isolated, with minimal financial resources. The average length of stay in a nursing home is measured in years, quite a contrast to the 7 to 10 day average for acute hospital stays. In light of recent public concern about nursing home conditions, administrators, boards of trustees, and involved families are leery of research projects, fearing abuse of patients. A researcher in the nursing home must approach investigational effort with caution and is obligated to satisfy the stringent criteria of the Institutional Review Board, as well as the Trustees and Administrators, who fear negative publicity for their facilities.

A corollary to this fear of abuse is the sentiment of many patients or their families that potential subjects have "suffered enough already," fearing that they are being used as "guinea pigs." These concerns reflect societal concern about the risks of research in the absence of perceived benefits to society at large, if not to the subject directly.

Residents have often been admitted to nursing homes at the request of, or under the persuasion of, family members, community health professionals, or community agencies, and may view themselves as "captives" of the institutions.

Because congregation of the elderly in nursing facilities forms a pool of subjects that may be viewed as easily accessible by researchers, care must be taken that they not be exploited by extensive involvement in research projects.

The diagnosis of dementia is often threatening to patients and their families. Though families who have accepted this illness in relatives may be eager for research on the disease, families confronted with a diagnosis they have not accepted will be further threatened by requests for research.

The Institutionalized Elderly

Before delineating other characteristics of the institutionalized elderly, it may be useful to examine the spectrum of patients to be found in nursing facilities. A minority of residents may be adults younger than the seventh decade suffering from strokes, malignancies, pre-senile Alzheimer's disease, or similarly handicapping conditions. However, the average age of most nursing home populations is considerably over the age of 65, usually between 80 and 85. These patients are indeed the "frail elderly."

The range of these patients includes:

1. Individuals with significant physical disabilities who retain full mental capacities
2. Individuals with mild or moderate dementia who may remain competent to consent for therapeutic or research procedures while simultaneously exhibiting incompetence in other spheres
3. Individuals with more severe dementia, whose competence is uncertain
4. Severely demented individuals with whom communication is impossible

Many traits hindering attempts to obtain informed consent from the nursing home population are shared with the elderly in the community.[2,4] However, these same characteristics have contributed to the very frailty that necessitated institutionalization of these patients and, as a result, tend to be more prominent in the nursing home setting.

Vision and hearing disturbances are not uncommon. Can the subject hear well enough to understand an explanation of the research procedure? Does the subject's vision permit perusal of a detailed informed consent form? At the very least, a researcher should provide forms with large type and use simple language readily understood by the average lay person.[4,7]

Communication problems also exist because of language barriers between patients and caretakers of different backgrounds. It is not infrequent for both the patient and the caretakers to be more fluent in languages other than English. In addition, limited education and poor vocabulary skills impede understanding of technical documents. Dysphasias or aphasias may competely hamper attempts to assess competence.

There appears to be proven evidence for impairment of cognitive functions with aging.[4,7,11,] Older subjects may require longer time to assimilate the contents of an informed consent request, and impaired judgment may prevent a reasoned decision. If recall is to be an essential element of informed consent, a stipulation that is open to debate, then the memory decline which occurs with age is another obstacle to the informed consent process.[12]

Mental fragility is manifested by alterations in mental state that may accompany changes in the environment, lack of sleep, emotional distress, and even minor illness. Even the psychological stress of mental status testing or interviews to assess comprehension of a proposed research project can be detrimental. A patient may thus be assessed competent to give consent one day and incompetent the next.

Elderly subjects are often on numerous medications, many of which may adversely affect mental function. In particular, institutionalized subjects frequently receive tranquilizers and may, indeed, be oversedated, an obvious barrier to comprehension.

Many of the elderly are fearful of signing documents. Although they may verbally give consent, they may refuse to sign consent forms.[4,5]

The process of institutionalization itself has an effect on the elderly individual's capacity to give informed consent. Cared for by a physician on the staff of the institution, the patient has often had to break longstanding ties with his or her primary physician in the community. This break may represent the severance of a trusted physician–patient relationship, thrusting the patient into a vulnerable position. Many patients seek to please their caretakers, either from fear of reprisal or because of the trust that they have developed for the staff. They may thus be subject to "subtle coercion," consenting to procedures for which they sense their caretaker's approval. One must thus question whether a researcher can simultaneously function as primary physician for his or her subjects, a situation not unlikely to occur in a geriatric facility.

The dependency promoted by institutionalization prompts some patients to seek assistance with decision-making. In our institution, we observed one researcher's attempts to obtain consent for a research protocol. Of 25 women determined to be competent to give consent, eight would not give consent until their decision had been approved by a trusted head nurse, a family member, or their primary care physician.

A double dilemma exists here: will the presence of a trusted family or staff member facilitate a subject's comprehension of the consent request, or will the patient instead view the trusted person's presence as encouragement to give consent? Of course, there are situations in which caretakers may actually be overprotective of their patients and attempt to dissuade them from participation in research. It is essential, indeed, that the patients' rights to participate in research is not denied by excessive precautions because of their vulnerability.[5,13]

Peer pressure is another factor to be considered. Discussion among residents on the wards or in the dining rooms can significantly influence the patients' responses to research protocols.

The lifestyle of the average nursing home resident tends to be inactive; involvement in a research protocol may be viewed as a means to interrupt the monotony of institutional living.[5]

The "aloneness" of many individuals, often single or widowed women with no children, fosters dependence on the institution. In addition, such "aloneness" with its accompanying depression may compel a patient to accept risks which others might consider with more caution.[4]

Finally, an element of altruism exists: patients who view themselves as severely incapacitated by their diseases may feel that participation in research gives purpose to their lives. Alternatively, such involvement may represent a form of sacrifice for the benefit of society. Family members of severely demented patients may share such sentiments.

Improved Care

This compendium of obstacles to informed consent is not intended to foster despair of ever performing research on the institutionalized dementia patient. Research is needed, not only to understand the pathophysiology of SDAT, but also to explore improved ways of caring for these patients.[4] The "teaching nursing home"[15] doubtless represents a key site for research on SDAT. Here patients experience skilled care in an environment promoting education, a setting conducive to research. The quality of health care provided in the "teaching nursing home" permits confidence in the accuracy of results that might otherwise be adversely affected by inadequate care.

Nevertheless, the issue of informed consent remains.[16] Can there be truly informed (let alone, educated) consent? This ideal is difficult enough to achieve with younger adults; it would seem virtually unattainable among the frail elderly. The informed consent process represents an attempt to regulate the physician–patient relationship, requiring disclosure of benefits and risks to the patient, based on the principle of autonomy, or self-determination[7,8]: alternatives must exist. Regardless of the process used to obtain consent, an underlying principle is the patient's faith in his or her caretakers. As Ingelfinger has stated, the potential subject is really protected by the "conscience and compassion of the investigator and his peers." Libow suggests, therefore, a concept of "informed trust," in which the physician–patient relationship promotes the patient's confidence that the physician's decision to include the patient in a research project is a moral one. Thus, this concept does not require full comprehension of the potential benefits and risks, although they must be discussed with the subject. One obvious caveat is the promotion of trust by an unprincipled researcher, who can thus encourage a subject to acquiesce to unwarranted risks for the sake of research.

The concept of "informed trust" does not solve the issue of how to obtain consent for research in severely demented patients.[2] Must consent by a surrogate be, in fact, truly informed? Can a surrogate give consent for research involving any, or more than minimal, risk?[6] Who is the legal surrogate for an elderly patient? Though the surrogate for a child or a retarded individual is readily identified as the parent, the surrogate for an elderly person is not well defined, particularly if there is no living spouse. The children or next-of-kin are often accepted as surrogate for therapeutic decisions. However, conflicts may arise as to which relative is actually the next-of-kin, and courts may not even accept the decision of a next-of-kin if the therapeutic measure was not judged to be lifesaving.[9] If physicians are to be guided by the courts, consent for research that has no direct benefit to the patient may not be obtained from surrogates. Such a stipulation will severely stifle research which may offer considerable benefit to society as a whole. The Penultimate Will, or its equivalent, represents a possible solution[18]: each individual, while still competent, can thus legally appoint a surrogate and inform that surrogate of his or her beliefs and wishes regarding care and research should the individual become incompetent. Though one could argue that an individual's values may change between the time the will is signed and the time he or she becomes incompetent, certainly a statement of the subject's own wishes should bear more significance than an interpretation of his or her values by a next-of-kin or a court-appointed surrogate. One is led to ponder the implications of requiring each competent patient to sign a Penultimate Will at the time of admission to an institution.

Summary

This paper outlines factors that contribute to the physical and psychological vulnerability of the institutionalized elderly. The "teaching nursing home" is advanced as the ideal setting for research on institutionalized dementia patients. The concept of "informed trust" is suggested as an alternative to truly informed consent. The final proposal is for the use of the Penultimate Will with instructions regarding research.

Notes and References

[1]H. K. Beecher, "Ethics and Clinical Research," *New England Journal of Medicine* 274, 1966, 1354–1360.

[2]S. Berkowitz, "Informed Consent, Research and the Elderly," *Gerontologist* 18, 1978, 237–243.

[3]R. N. Butler, "Protection of Elderly Research Subjects," *Clinical Research* 28, February 1980, 3–5.

[4]M. P. Lawton, "Psychological Vulnerability of Elderly Subjects," in NIA Conference on the Protection of Elderly Research Subjects, DHEW 79–1801, pp. 6–13.

[5]J. L. Makarushka and R. D. McDonald, "Informed Consent, Research, and Geriatric Patients: The Responsibility of Institutional Review Committees," *Gerontologist* 19, 1979, 61–66.

[6]NIA Conference on the Protection of Elderly Research Subjects, NIA, Bethesda, MD, 1977, DHEW 79–1801.

[7]R. M. Ratzan, "Being Old Makes You Different: The Ethics of Research with Elderly Subjects," *Hastings Center Report* 10, October 1980, 32–42.

[8]W. T. Reich, "Ethical Issues Related to Research Involving Elderly Subjects," *Gerontologist* 18, 1978, 326–337.

[9]A. J. Rosoff, "Informed Consent: A Guide for Health Care Providers," Rockville, MD, Aspen Systems, 1981, 522 pp.

[10]J. B. Wales and D. L. Trybig, "Recent Legislative Trends Toward Protection of Human Subjects: Implications for Gerontologists," *Gerontologist* 18, 1978, 244–49.

[11]H. A. Taub, "Informed Consent, Memory and Age," *Gerontologist* 20, 1980, 686–690.

[12]R. Miller and H. S. Willner, "The Two-part Consent Form: A Suggestion for Promoting Free and Informed Consent," *New England Journal of Medicine* 290, 1974, 964–66.

[13]R. M. Ratzan, "The Experiment That Wasn't: A Case Report on Clinical Geriatric Research," *Gerontologist* 21, 1981, 297–302.

[14]C. K. Cassel and A. L. Jameton, "Dementia in the Elderly: An Analysis of Medical Responsibility," *Annals of Internal Medicine* 94, 1981, 802–807.

[15]R. N. Butler, "The Teaching Nursing Home," *JAMA* 245, 1981, 1435–1437.

[16]F. J. Ingelfinger, "Informed (but uneducated) Consent," *New England Journal of Medicine* 287, 1972, 465–466.

[17]L. J. Miller, "Informed Consent," I. *JAMA* 244, 1980, 2100–2103. II. Ibid, 2347–2350. III. Ibid, 2556–2558. IV. Ibid, 2661–2662.

[18]R. Zicklin and L. S. Libow, The Penultimate Will, *NY State Bar J* 1975, 31–33, 63–65.

Legal Issues in Research on Institutionalized Demented Patients

Nancy Neveloff Dubler

Introduction

Illnesses that diminish, and may extinguish, mental capacity are feared, even in our cancerphobic society, more than the ultimate ravages of metastatic diseases. Certain dementias inexorably erode mental abilities, destroy sentience and humanity, and transform a human being into an object–organism. Because these diseases are so horrible and so threaten the humanness of those who contract them, research to uncover effective treatment must be considered, despite the apparent problems with individual informed consent.

The task of this essay is to analyze the legal issues that arise in including demented institutionalized patients who cannot, under any proposed standard, provide informed consent in research protocols. The goal is to propose a system of legal protections and procedures that could permit potential benefits of research to individual patients and to society, while simultaneously protecting against abuse and misuse of afflicted patients.

The process of research, once an unquestioned good, is now suspect in part because researchers have been perceived to disregard the rights[1] of involved subjects. Despite the currency of this opinion, one should consider that:

> *Although the Nazi atrocities exposed at Nuremberg awakened public interest in the regulation of experimentation using human subjects, one must remember that what makes this subject difficult is that most medical experiments are not atrocities, but sincere efforts to improve the well-being of the species and often the individual subject by increasing our knowledge and our weapons against disease. Exces-*

> *sive squeamishness about hypothetical horrors is Scylla to the Charybdis of exploitation of the individual well-being for some asserted higher group good.*[2]

The problem with the task at hand is that it must address previously evolved concepts of informed consent and sets of regulations. Informed consent in cognitively capable persons acts, at least in theory, to empower the individual patient and to limit the actions of others. Informed consent as a matter of law reflects values of individual liberty and self-determination that become meaningless absent coherent thought processes. It may thus represent the wrong set of values to be applied to this discussion. If there is in fact some "higher group good" to be sought in research on the origins and treatment of SDAT, informed consent may not provide the necessary mediating principle.

Rights for Those of Diminished or Developing Capacity

The law has not settled beyond question or debate that persons do not, merely by reason of age—either advanced or immature, or by reason of curtailed physical freedom (i.e., jail, prison, mental institution, or nursing home)—relinquish certain fundamental rights. Among the rights that accrue to all and that survive, in some form, the process of confinement are basic rights to self-determination and autonomy in medical contexts[3] and the constitutional right to privacy.[4]

This growth of rights for the segregated and hidden—and unfortunately often the forgotten—has developed in the last two decades. In prisons, for example, where rights are diminished by statute and by conventions of coercion, the federal courts and the Supreme Court had traditionally adopted a "hands-off" policy, citing the complex nature of these institutions (the difficulties of administration and the intractability of problems) that made them particularly inappropriate for judicial decree. Despite the evident truth of the perceptions supporting this "hands-off" doctrine, the Supreme Court finally decreed that prison administrators must justify previously unquestioned policies when challenged and must demonstrate the necessity for any deprivation of basic and protected rights. The court held that, "When a prison regulation or practice offends a fundamental constitutional guarantee, federal courts will discharge their duty to protect constitutional rights."[5] Since that opinion in 1973 a panoply of rights have emerged that attempt to protect prisoners from unwarranted punishment and abuse in general, and from deprivation of medical care in particular.[6] Other cases established that prisoners must consent to and may refuse treatment, even though they have clearly diminished voluntary ability to exercise rights of self-determination and autonomy.[7]

Children whose rights, like their judgment—which is assumed to underlie the exercise of rights—have not fully matured, also have been guaranteed liberties of action in quasicriminal proceedings,[8] in attempts to secure abortions,[9] and in support of continued life over the attempt of parents to refuse care and treatment on their behalf.[10]

The mentally institutionalized and retarded have also been the focus of legal actions designed to establish basic rights to treatment,[11] a qualified right to refuse treatment,[12] and the right not to be "warehoused" and discarded.[13]

In all of the instances above individuals with diminished capacities of self-determination and autonomy, either because of immature or impaired judgment—children and the mentally infirm—or by virtue of the nature of their confinement—the criminally sentenced—have been provided with both doctrine, and with specifically designed protective procedures to insure support for their personal integrity. Protections for the incapacitated, incapable, infirm, and incarcerated especially in congregate settings is now the norm.

Institutionalized populations, given recent history, are likely to be the subject of special judicial scrutiny if it appears that their rights are in jeopardy. The segregated possess the rights of the free, albeit in slightly altered forms. Thus if we are to vary the general requirements for informed consent to research, justifications must be articulated for varying the normal for new approaches will protect patients from abuse and researchers from later liability.

Research in General

It is now established by a complex of Federal Regulations[14] that before biomedical experimentation with human beings may be conducted, the protocol must be approved by a properly constituted Institutional Review Board (IRB) that must determine:

1. That there is a positive risk-benefit ratio—that is, that the risks are sufficiently "outweighed by the sum of the benefit to the subject and the importance of the knowledge to be gained" so that the subject may be permitted to weigh the personal risks and benefits
2. That the subjects' well-being will be adequately protected during the course of the research
3. That the informed consent of the subject will be obtained[15]

If these prerequisites are met, the researcher is permitted to present the following to an individual subject: a description of the protocol; a statement of the risks and benefits; the assurance that participation will not be continued beyond the wishes of the patient; and a statement describing compensation for any negative outcomes of the research. The patient may then consent to or refuse participation.[16] This process assumes cognitive ability. It does not require rational decision-making and does not demand wisdom; it does assume intellectual capability.

The questions relevant to this discussion are:

1. Are these patients, particularly institutionalized patients, who by virtue of dementing illness, are incapable of providing informed consent to participation in research protocols but who may, nonetheless, be included in research?
2. If there are such patients, what structure for substituted consent would suffice and what additional protections would be required to provide adequate protection for patients?[17]

These questions assume that the individuals under discussion are not and will not be capable, under any definition of informed consent, to participate in the usual decision-making process.[18]

If the present proposed regulatory structure governing research involving those institutionalized as mentally infirm is used as a guide, research would be permissible on a patient who could not provide informed consent if the physician obtained consent from the patient's proxy or, in the language of the Federal Regulations, the patient's "legally authorized representative." The legally authorized representative becomes critical when the physician suspects that the usual rules governing informed consent, i.e., the requirement of an "adult individual of sound mind," do not apply. The Federal Regulations do not define who is a "legally authorized representative," thus referring the matter to applicable state law. Unfortunately, most states have neglected to provide an adequate definition of the concept or the process by which the label attaches.[20] Thus, these regulations and the exception they provide offer little concrete guidance. Moreover, I will argue that a legally authorized representative or guardian, if not specifically empowered to consider research issues, need not and probably should not be permitted solitary approval.

Some kinds of research exist that could advance the understanding of the disease or the treatment of these patient(s) and may be performed without requiring or securing informed consent. These are protocols in which the underlying interests protected by the requirement of informed consent are not at issue, i.e., there is no threat to the autonomous exercise of rights, no possibility of harm, no danger of breach of confidentiality, and no elements (either in design or projected outcome) to which a reasonable person could be expected to object.[21]

Such experiments might include an analysis of blood or urine, drawn for unrelated clinical care purposes that would otherwise be discarded. Certain retrospective record reviews might also fall in this category. Such protocols yield some helpful data and in no way threaten the protected interests of the patient, assuming of course that the implementation of the protocol in no way modified the previously existing patterns of care. Protocols of the type that elicit information about the course and social implications of a disease through interviews with the families of patients, would require obtaining the informed consent of the persons interviewed.

Finally, observational research could arguably fall within the category of protocols that require no informed consent. The contrary position, however, seems weighty: systematic observation of the behavior of an incompetent patient, behavior that would in all likelihood be a profound embarrassment to the same individual if competent, seems to constitute precisely the assault on personal integrity that the requirement of informed consent is designed to mediate.

Assuming, therefore, that there are protocols for which informed consent is not required and which can thus proceed merely with IRB approval, three categories remain:

1. Non-intrusive, non-invasive data collection and observation. Invasive research with some possibility of direct therapeutic benefit to the patient.
2. Invasive research with some possibility of direct therapeutic benefit to the patient.
3. Invasive research with no possible therapeutic benefit.

This discussion will proceed to examine whether in the above situations an adequate substitute for the voluntary and informed consent of the patient can be devised.

The Kaimowitz Case

The leading, and until recently, almost the only case,[22] involving legal issues in human experimentation is the case of *Kaimowitz v. Dept. of Mental Health for State of Michigan.*[23] This case preceded the deliberations of the National Commission for the Protection of Human Subjects and dealt with the question of the permissibility of psychosurgery on an involuntarily institutionalized patient. The court characterized this surgical intervention as dangerous, intrusive, irreversible, and of uncertain benefit to the patient and society.[24] Given the initial characterization it is not surprising that the court found that prerequisite to an intrusion "upon the body of a person must be full, adequate and informed consent,"[25] and that involuntarily confined mental patients who live in "inherently coercive institutional" environments are not capable of providing this voluntary informed consent.[26]

Two elements of the case are particularly interesting. First, the defendants produced a document from the patient purporting to grant informed consent. They further produced evidence that the parents supported that consent. In addition, the testimony showed a prior scrutiny of the underlying protocol and of the validity of the consent by two specifically convened committees. One of these committees addressed the scientific validity of the experiment and the other addressed the specific adequacy of the process and substance of the consent. Despite this charade of an involuntarily detained mental patient, these procedures were found not to suffice and probably could never do so.

Basic to the analysis of the court was its evaluation of the inherently coercive nature of the kind of institution in which the patient was detained. This would argue not for a wholesale exclusion of all institutions, but rather for the adequate individual examination of any institution considered as a setting for research, to determine whether or not it is indeed coercive, although the court in the Kaimowitz case did not so limit its language.

The Institution of the Nursing Home

Most severely demented patients are housed in nursing homes. The term "nursing home" refers both to Skilled Nursing Facilities (SNF) and Intermediate Care Facilities (ICF) both defined by federal statute.[27] The first is eligible for reimbursement of services from both Medicare and Medicaid while the second is eligible only under Medicaid. Both types of institutions are subject to federal and state regulations governing facilities, staffing, and services provided.[28] The inability of states, entrusted with the enforcement of federal regulations to ensure protection for patients and an acceptable quality of care, has been documented by various state and federal investigations, by scholars, and by popular press.[29]

Nursing homes stand as the latest in a series of institutions beginning with the poorhouse, the prison, and the asylum that have been devised by society to deal with the "decrepit outsider."[30] Historically, the investigation of these institutions has uncovered individuals in desperate positions, "stripped of power and desolate of dignity."[31] These institutions place a premium on compliant, nondisruptive, quiet behavior; the staff accolade for the nursing home patient is "good girl" or "good boy"—meaning neither disruptive, dirty, demanding, or otherwise bothersome.

All total institutions are at risk of becoming inherently coercive settings. However, the nursing home has an additional motive to support tyranny and deprivation: the motive of profit. As services are cut and staff rotations curtailed, the percentage of profit on maintenance contracts with individual families or with the state rises. The maximization of profit, as evidenced by our fee-for-service system of medical care, has often been charged to be incompatible with the provision of justly distributed decent care.

In response to demonstrated violations of the rights and liberties of nursing home patients (even competent nursing home patients) the United States Department of Health and Human Services had proposed a new section of regulations governing the conditions for federal reimbursement of nursing home institutions and providing new protections for the rights of their patients.[32] These regulations include, among others, protections for the exercise of rights without restraint, interference, coercion, discrimination, or reprisal from a facility; the exercise of an incompetent patient's rights by a legal guardian or next-of-kin; notice to patients of facility policies and procedures; freedom of association with other persons; access to desired visitors; limitations upon involuntary patient transfer. These regulations not only describe the variety and dimensions of abuse to which they attempt to respond, but if adopted, would have created a new legal basis for protecting institutionalized incompetent patients.

Nursing homes have often demonstrated little regard for patient autonomy and dignity, and have often operated in disregard of the rights, prerogatives, and actual care needs of the persons they are designed to serve. This history does not rule out all research, but argues for an individual determination that the specific setting is in fact caring, supportive, and nonabusive.

Privacy Rights and the Substituted Judgment Doctrine

Two recent cases, *Rennie v. Klein* and *Rogers v. Oken,*[33] explore the right of possibly incompetent patients to make decisions and comment on the underlying constitutional rights that support patient autonomy and protect even idiosyncratic decision-making. In both cases, the right to "privacy" has been extended to provide a basis for a qualified right to refuse psythotropic medications for those confined in mental institutions. The right to privacy, as defined, clearly encompasses the interests protected by the doctrine of informed consent:". . . the areas of autonomy protected by the right to privacy are particularly personal and fundamental; it is clear that decisions concerning medication that affect bodily integrity and the mental processes are encompassed by that right."[34] This right to privacy, as in previous judicial formulations, was not found to be absolute, but rather to be subject to regulations as the interests of the state are balanced against concepts of autonomy.[35]

The right of privacy does in fact battle continuously against increasingly compelling state interests. In the matter of abortion the state's interest in the preservation of life outweighs the right of privacy in the third trimester, and then permits the prohibition of abortion. A variation of this same state interest in life permits overriding the right of the mentally infirm to refuse medication, and to permit the imposition of this state right against articulated individual expression, although specific procedural safeguards are provided.

Similarly, it could be argued that the state has a stake in the expansion of medical knowledge and technique, and that the state's interest in protecting its citizens from the ravages of dementing illness could, under certain conditions and with stipulated protections, outweigh a patient's right to privacy under this line of reasoning.

Substituted judgment—that process which seeks to permit the right of incompetent patients to "refuse" care and treatment—rather than a privacy analysis, provides perhaps the closest analogy to the right of a "consenter" to enroll an incompetent patient in a research protocol. The quality of the judgment in refusing care for another and the immediate and direct harm to life that such decisions often produce is most akin to a third party substituted consent for participation in research with the possibility of harm that it too must acknowledge if the protocol demands. Both judgments possibly threaten the incompetent's interest, either in life or in the non-subjection to risk.

In arguing that substituted judgment provides a basis for permitting incompetents to donate organs to another, one comment states that since incompetents are treated as persons in some important respects,

> *Consistency requires that, when questions arise concerning their treatment in particular situations they also be treated as persons with wants and preferences. By failing to treat them as we treat competent persons in similar situations ascertaining and respecting their lawful choices we might undercut respect for the incompetent person in other situations and eventually diminish respect for all persons.*[36]

This analysis would argue for permitting some third parties, under some circumstances, to permit an incompetent to participate in research.

The legal roots of the substituted judgment lie in concepts of property law and the tests devised to permit distribution of the assets of an incompetent. Two tests were fashioned as guides: one subjective—what this particular person would do faced with this particular choice—and one objective—what the reasonable, rational person would do. The second argues that factors can be weighed to reflect what a competent person with the "characteristic tastes, preferences, history and prospects"[37] of the incompetent would be, thus maximizing a pattern of interest, a pattern of want, or a prior, personal, idiosyncratic history. It can be argued that it represents the greatest respect for persons in general to permit a person when incompetent to continue to exercise, through the judgment of others, those options that would have been available had incompetence not intervened. Thus, permitting a substituted judgment under this argument represents the maximal support for the concept of autonomy.

Similarly it can be argued that rules permitting third party permission in the context of nonmedical interventions serve to permit the exercise of "choices while at the same time protecting incompetents from undue harm."[38] By this analysis third party permission should not be allowed if it increases "in anything more than the most minimal fashion, the ratio of risks to benefits."[39] Furthermore, it can only proceed if the substitute decision-makers have the following characteristics:

1. No conflict of interest with the incompetent
2. An ability to participate in a vigorous, informed, and conscientious manner in the decision
3. An ability to remain a vigorous advocate of the incompetent's interest maintaining control of decision making throughout the intervention.[40]

The doctrine of substituted judgment has been pivotal in those cases dealing with questions of third party attempts to withdraw or to refuse medical care for an incompetent. The cases involving patients Quinlan, Saikewicz, and Dinnerstein[41] struggle to develop a theory and process of substituted judgment that could permit the rights of competent adults—including the right to refuse care and treatment even if the results be death—to be applied to incompetent persons. The Quinlan case involved a once-vibrant young woman now in a chronic-vegetative condition, with no reasonable possibility of ever returning to a cognitive or sapient state, who was being maintained on a respirator; her father petitioned to be appointed guardian in order to be empowered to turn off the life-support systems. The Saikewicz case involved an institutionalized congenitally profoundly retarded 67-year-old man with leukemia presented with the possibility of life-prolonging chemotherapy; the superintendent of his institution sued to determine who had the right to accept or to refuse treatment, and on what grounds. The case of Shirley Dinnerstein involved a 67-year-old woman suffering from end stage Alzheimer's disease; it questioned whether, in the event of a cardiac arrest, she needed to be resuscitated, and who should be empowered to decide the issue.

In these three cases, the courts located the basis for permitting substituted refusal of lifesaving or life-prolonging treatment in individual interests of autonomy, self-determination, and the right to privacy. All agreed that these rights survive incompetence and are best safeguarded by permitting independent third-party extension. A third-party decider (either an individual, the court, or a physician with the counsel of family) exercises the basic right of refusal of care on the "subjective" basis of what this incompetent patient would decide if he or she were competent to do so. The doctrine thus appears to be uniform in application; in fact, there are great differences.

In the case of Karen Quinlan, there is some logic in trying to determine from her prior statements, behavior, actions, and preferences what her desire would be. For Mr. Saikewicz, it clearly pushes the doctrine of subjective substituted judgment beyond any reasonable bounds. As Mr. Saikewicz was indeed never competent to decide and thus never competent to state preferences it is an objective judgment (i.e., what one as a reasonable person could decide) presented in subjective garb. In the Quinlan case the judgment was to be executed by a specially appointed guardian, after a Bioethics Committee had confirmed the prognosis of no possible return to sapient existence (and parenthetically delimited the liability issues). In Saikewicz the court itself exercised this judgment, after a full adversary proceeding, with the participation of a guardian-ad-litem.

The case of Shirley Dinnerstein, although ostensibly related to the others, is really quite separate; it was characterized as a case involving the natural death of the terminally ill patient that presented the question of appropriate measures "to ease the imminent passing of an irreversibly, terminally ill patient."[42] Dinnerstein reflects normative medical practice under which extraordinary medical efforts are not made for clearly terminal comatose patients. In such cases the necessity to consent to or to refuse treatment does not arise; the treatment is not a viable alternative. The Dinnerstein court distinguished Saikewicz as a case that involved not the mere postponement of the act of dying, but the possibility of a remission of symptoms enabling a return to a "normally functioning integrated existence."[43] In Dinnerstein the private world of medical decision-making is properly protected against intrusions of a public process.

Contrast these cases with *The Matter of Spring,*[44] which involved a patient with chronic organic brain syndrome, irreversibly demented, who also had end stage kidney disease. The case raised the issues of the ability of a guardian to order the termination of life-sustaining hemodialysis treatments. Upon petition of the guardian to the court, the court appointed a guardian-ad-litem who opposed the allowance of a petition to terminate dialysis and upon the granting of that petition appealed the order. The appeals court found that the patient would wish to have dialysis treatments discontinued. This finding was based not on any explicit statements by the patient, but rather on the consideration of an eclectic list of factors including: prior life style and patterns of action contrasted with the present decrepit state of existence; effect of the treatments and their side effects; wishes of the family; and counsel of the attending physician. In balance the court affirmed an order requiring the temporary guardian to terminate lifesaving treatment.[45] On further appeal the Massachusetts high court agreed that the patient would probably have discontinued treatment if able to do so. It reaffirmed, however, that a question of withdrawal of treatment from an incompetent patient, once presented to a court, must be decided by judicial process—the decision cannot be delegated to relative or caregiver.[46]

Finally, consider a recent New York case[47] that involved Brother Joseph Fox, an 83-year-old member of the Society of Mary, who during surgery suffered a cardiac arrest that left him on a respirator in a vegetative state. The accompanying case involved John Storar, a 52-year-old man with a mental age of 18 months who was afflicted with bladder cancer and sustained by regular blood transfusions. The court permitted the discontinuation of life supports for Brother Fox, stating that Brother Fox had "made the decision for himself before he became incompetent,"[48] and commented that permission to terminate treatment was designed to "give effect to an individual's right (that of self-determination and the right to refuse) by carrying out his stated intentions."[49] The court, applying a standard of clear and convincing evidence, found that Brother Fox had explicitly declared that if in a vegetative state, he would not want his life sustained by a respirator.

The court then distinguished the case of Storar as one involving an individual who, since he was never competent, could never give a clear indication of his wishes and desires in any particular circumstance. By analogy to decisions about the treatment of children where, said the court, a parent, no matter how well-intentioned, may not deprive a child of lifesaving treatment, the court declared it could not permit an incompetent patient to "choose" death because someone, "even someone as close as parent or sibling, feels that this is best for one with an incurable disease."[50]

As a result of these cases, it is arguable in New York State that there is no doctrine of substituted judgment. For those congenitally incompetent no one may decide to refuse treatment; for those once competent, neither of the traditional subjective or objective tests will suffice, but only proof of explicit statements demonstrating specific desire to discontinue treatment.

Third-party refusal of care for an incompetent patient shares certain characteristics with possible third-party consent for inclusion of incompetent patients in research protocols. In both the legally protected interests of the patient in privacy, self-determination and the exercise of choice are threatened by incompetency. In both, the state has a substantial series of interests in ensuring that the legal rights of the incompetent be protected, the right to continued life be supported, and the integrity of medical practice and decision-making be secured. In both, the fabric of agreement that usually supports decision-making in a medical context is torn asunder. Whereas my suspicion is that Eichner sets too rigorous a standard in "termination of treatment" cases, it might be quite reasonable as a standard governing experimentation without consent. In these latter instances where the patient may be used as a means to a societal end—i.e., increase in knowledge—a prior explicit consent allays fears of trampling on rights. In the former, however, where termination is supposedly in the best interest of the patient and not in support of a "good" of others, the requirement of explicit empowerment may lead to unnecessary prolongation of suffering.

Substituted Judgment as the Basis for Consent to Research

Is there a "right to participate in research" or to "choose" research which, by analogy to a right to refuse treatment, could be protected by appointment of a third-party decider or by a judicial decision maker? Certainly, in the case of research with possible therapeutic outcome, one could argue that the ability to choose such a course should survive incompetency. If so, a judicial process with the appointment of a guardian-ad-litem to argue the wishes, benefits, and interests of the patient (Saikewicz and Spring) may need to surround the exercise of such judgment. Other models (Dinnerstein and Quinlan) would be inadequately public, and thus inherently inappropriate to many experimental, as opposed to treatment, contexts. Were one, however, bound by the most stringent standard for substituted judgment (Eichner), a substituted consent would be precluded for those not previously competent and would require, for those who had been competent adults, a prior explicit statement of direction.

This sort of rigorous standard is not inapproriate as the basis for permitting a third party consent that would carve out an exception to usual legal and medical research norms.

If we consider again the major (possible) variations of research:

1. Non-intrusive, non-invasive data collection and observation
2. Invasive research with some possibility of direct therapeutic benefit to the patient
3. Invasive research with no possible therapeutic benefit to the patient, and if we consider also the appropriate characteristics for a substitute decider which include:
   a. No conflict of interest with the incompetent
   b. An ability to participate in a vigorous, informed, and conscientious manner in the decision

c. An ability to remain a vigorous advocate of the incompetent's interest maintaining control of decision-making throughout the intervention, and

if, finally, we consider the setting: nursing with a history of abuse of patient's medical, physical, and legal rights, then certain solutions become possible.

For non-intrusive, non-invasive data collection, a substituted consent of the guardian might be appropriate:

1. If the guardian possessed the characteristics of independence, commitment, and supervision described above[51]
2. If a neutral fact-finder[52] has determined that the institution is in fact noncoercive, and that Federal and state regulations governing the care and treatment of patients have been complied with
3. If the research has been reviewed and approved by a properly constituted IRB that has determined that it will not subject the individual to undue harm or more than "insubstantial risk."

Invasive research with no possible therapeutic benefit, that is, the use of incompetent institutionalized persons as cannon-fodder for a war of unrelated research, is somehow repugnant to concepts of equity and fairness and violates personal integrity unless there has previously been an indication of acceptance.

Is there, finally, a role for third-party deciders in consent to research that carries some risk and some possible therapeutic effect? Given the history of the abuse of research, the nature of some nursing institutions, and the oft-stated regulations designed to avoid the excesses of research, I would argue that only a public judicial process, with a guardian-ad-litem, subject to adversary presentation, would be adequate to insure appropriate protection and to avoid giving undue preference to possibly conflicting interests. This solution presents obvious problems for the investigator who must secure the time, financial support, and access to legal resources. It also presents an additional burden to an already swamped judicial system with little expertise in biomedical decision making. However, all of the alternatives from private committee review to overgeneralized government, bureaucratic assurance of the rights of subjects are insufficient. All are potentially ineffective, subject as they are to the operation of democratic process, the overriding acceptance of co-opting research norms, and budgetary cutbacks. No private system can assure protection; only a public forum governed by the adversary process can protect the "decrepit outsider."

Thus non-intrusive, non-invasive research with the additional element of no more than "insubstantial" risk or inconvenience could be permitted by third-party deciders. Adequate examination by an IRB would be necessary, as would the consent of an advocate appointed for the patient, and examination of the particular institution to insure it is in fact noncoercive. Invasive research, however, must seek judicial permission.

An Alternative: Consent Prior to Dementia

One of the difficulties encountered in devising rules for obtaining the informed consent of Alzheimer's patients is the understandable, but nonetheless oppressive, silence that surrounds the familial and professional care of these patients. The concept of declining competence and impending dementia is more terrifying for many than impending death. The fact that one will continue to exist in form but with totally altered substance, appearing without dignity and deportment, is a terrifying prospect for patients and an unmentionable fact for physicians.

It is my strong impression (gathered in discussions with physicians, nurses, and ancillary personnel who work with incompetent patients and families) that it is more difficult for a health staff to discuss with a patient that patient's impending incompetence than it is to discuss with a patient the natural course of a terminal illness.

Over the last two decades, extraordinary strides have been made in helping physicians and patients to deal in a straightforward, honest, and useful manner with impending death and the legal issues and treatment decisions that may surround it.

Physicians can discuss with patients who are seriously burned, for example, that their survival, given age, and the dimension of the burn, is unprecedented, and thus can, while the patient is still lucid, determine the patient's choice between a full therapeutic regime or ordinary care.[51] Similar discussions appear not to be held with those patients who will not possibly or probably but, certainly according to best medical evidence, be demented.

The open and frank discussion with patient would permit a number of solutions to the present problem. It would permit the patient to make an explicit statement that would satisfy the requirements of the most stringent (i.e., the Eichner) test for substituted judgment. It would be possible for a patient to say, in the context of a discussion about the course of the disease "If I am ever totally incompetent and there is the possibility of a new drug, process or procedure that would help to reverse my decline, I would want to participate in its trial." More particularly, however, it might permit the development of state statutes providing for substituted consent for participation in research protocols either by appointing a substituted authority, an agent—the Michigan model[55]—or by development of governing language such as the California Natural Death Act.[56]

In the Michigan model, an adult competent person appoints an agent who is empowered to accept or refuse medical treatment and who can prevail over any guardian appointed by a probate court. The statute could be drafted to permit appointment of one with explicit power to consent to research, even nontherapeutic research (the analogous acceptance of risk).

The above would be preferable to an extension of the model developed by California in 1976, known as the Natural Death Act. That statute requires that two physicians diagnose a terminal condition or illness, assure the patient that the application of life-sustaining procedures would serve only to artificially "prolong the moment of death"[57] and support the statements of the patient that if "my death is imminent whether or not life-sustaining procedures are utilized, I direct that such procedures be withheld or withdrawn and that I be permitted to die naturally."[58] The statute has been criticized as being too complex and convoluted to be generally helpful.[59] However, even if not appropriately complied with, and thus not binding, it would be excellent evidence of individual preference in action based on previously discussed doctrines of substituted judgment.

The concept of a "Living Will" has not received extensive state support to date primarily because of the difficulty of drafting language which will precisely reflect situations in which treatments are to be withdrawn or discontinued.[60] Research protocols—written documents accompanied by translations into lay language in their informed consent sections—should meet this objection as they define a situation with precision. Patients in the early stages of disease can be asked for consent for long-term protocols. The process of medicine is such that most decisions to withhold or withdraw treatment never leave the private confines of a tertiary care institution. Research cannot be so. Protocols must be submitted in advance, reviewed by internal review boards, be accessible to those engaged in the research, and survive the public scrutiny that the publication of results mandates. Research is essentially public and thus different from the ethically complex, private, but essentially individualistic, system of patient care. Therefore, despite the problems of drafting a statement that would empower an individual, diagnosed with inexorably declining incompetency to authorize later participation in a research protocol, the state of the law and the public nature of research argue for the attempt.

There is, of course, an even more simple solution—a general empowerment for research. Providers can discuss with patients their diagnosis and the probable course of their illness, and, while those patients are competent and uninstitutionalized, permit them to consider the risks and benefits of a research in general and to execute an adequate document, such as a "durable power of attorney," permitting some one to exercise the patient's "right to participate in research" once the patient can no longer do so. The cost of such a procedure will be a delay in possible advances in medical science and treatment of illness. The benefit of such a procedure is that it avoids the complex dilemmas that surround the attempt to include incompetent institutionalized patients in protocols.

Even for these patients, however, some advocate must be appointed to examine the institution in which the research will be conducted and to provide ongoing scrutiny of the process, including the right to refuse continued participation despite an earlier empowerment.

Possible Liability

There is a final reason for urging that either explicit, public, judicial solutions or statutory solutions providing clear extended individual authority be sought before permitting the inclusion of patients with irreversible dementia in research protocols. Incompetent patients in nursing homes represent preeminently the genuinely "helpless," for "individuals who suffer in various degrees from senility, confusion, paralysis or other conditions (. . . [are] . . .) unable to express or communicate a request for help, or even to comprehend the nature of their predicament."[61] Legal representation of these people raises significant questions about the traditional norms of attorney–client behavior and the principles of legal ethics that govern lawyer's conduct. Some argue that in regard to the helpless, the ethical principles that prohibit solicitation and that require client consent are not as absolute as they might at first appear. For the helpless, the attorney (by analogy to a physician with a patient in need of emergency care who cannot communicate) may infer consent to representation and defense. A three-pronged approach is proposed for attorneys attempting to identify situations where unilateral intervention is appropriate: (1) What is the extent of the client's helplessness? (2) Which are the legal interests at stake? and (3) What is the availability of an impartial decision-maker?[62] Certainly, patients with irreversible, advanced dementia would qualify as helpless. The possible nonconsentual inclusion of those persons in a research protocol (given the definitive nature of federal regulation and the existence of extensive case law defining the rights to self-determination, autonomy, and privacy) presents an immediate risk of irreparable harm to the helpless person's vital legal interest.

The mere inclusion of that individual in a research endeavor of any kind is an immediate violation of pre-existing rights to autonomy and privacy. Moreover, the existence of judicial precedent in determining the adequacy of substituted judgment certainly provides the availability of an impartial decision maker. Thus, it is not inconceivable that attorneys, shocked by the enrollment of these helpless persons in scientific studies, could, on their own initiative, seek to represent those persons in vindication of their rights.[63]

It is by now well-accepted that doctrines of equity demand that before special populations with diminished capacity for consent be used in research, all efforts should be extended to assure that consent be obtained that is appropriate, adult, competent, and informed. Individuals enter nursing homes for one of two principle reasons: either they need a level of care, both physical and emotional that they cannot provide for themselves and which their families either cannot or will not provide for them; or, given the institutional bias in public funding, they are incapable of maintaining economic integrity except in a nursing home. In either case the isolation and despair that characterizes these people require the support of a congregate facility. Depression or deterioration has often made them unable to cope. Disorientation has frequently rendered them incapable of managing in a modern world. Many of these patients represent precisely the "decrepit outsider" or "helpless" whom principles of equity and the forces of law declare must be protected.

Notes and References

[1]Hunt, *A Fraud that Shook the World of Science,* N.Y. Times, November 1, 1981, Section 36 (Magazine), 42.

Gup and Neumann, *Experimental Drugs: Death in the Search for Cures,* Washington Post, October 18, 1981, Section 1, 1, col. 1 (and the articles following).

[2]W. Wadlington, J. R. Waltz, and R. B. Dworkin, *Cases and Materials on Law and Medicine 1010 D.I* (New York: Foundation Press, 1980).

[3]Schloendorff v. the Society of New York Hospital, 211 NY 125 (1914).

[4]The right to privacy was first enunciated in cases establishing a right to use contraceptives, Griswold v. Connecticut, 381 US 479 (1965); and Roe v. Wade, 410 US 113 (1973); Doe v. Bolton, 410 US 179 (1973); extended clearly to minors in Bellotti v. Baird, 443 US 622 (1979); Planned Parenthood of Central Missouri v. Danforth, 428 US 52 (1976); and extended to the institutionalized in Rennie V. Klein, 476 F. Supp. 1294 (D.N.J. 1979); Rogers v. Okin, 478 F. Supp. 1342 (D. Mass. 1979); Mills v. Rogers, 454 US 936, 1136 (1981).

[5]Procunier v. Martinez, 416 US 396, 405 (1973).

[6]Estelle v. Gamble, 429 US 97 (1976), holding that "deliberate indifference to serious medical needs of prisoners constitutes the 'unnecessary and wanton infliction of pain' proscribed by the Eighth Amendment," (at 104). *See e.g.*: Winner, E., An introduction to the constitutional law of prison medical care. *Journal of Prison Health,* 1981, 1, 67; Dubler, N. The prisoner's right to medical care. *The Hastings Center Report,* October 1979, 7; Singer, R. Providing mental health services for jail inmates: legal perspectives. *Journal of Prison Health,* 1981, 2, 105; Ney, S. Prison overcrowding after Rhodes v. Chapman, *Journal of Prison Health,* 1982, 1, 5; Dubler, Jail and prison health care standards: a determination of need without reference to want or desire. In R. Bayer, A. Caplan, N. Daniels (Eds.), *In search of equity: health needs and the health care system* (New York: Plenum Press, 1983); Dubler, N. Commentary. *Man and Medicine,* 1980, 2, 87.

[7]The right to refuse care and treatment in prison is qualified by fears that denial of care will masquerade as refusal and by a commitment to ensuring that the inmate escapes neither to the outside world or to the next. It is also fettered by necessities of prison administration and by the peculiar public health issues presented by populations living in closed, closely-confined physical spaces. *See* Runnels v. Rosendale, 499 F. 2d 733 (9th

Cir. 1974); Commissioner of Corrections v. Myers, 399 NE 2d 452 (Mass. 1979).

The right of prisoners to consent to participation in medical research is examined in "Research Involving Prisoners," Appendix to Report and Recommendations, National Commission for the Protection of Human Subjects of Biomedical and Behavioral Research, DHEW Pub. No. (05) 76–132, 1966; and is addressed in 45 C.F.R. 46, 301–306. Because prisoners may be under constraints that could affect their ability to make a truly voluntary and uncoerced decision whether or not to participate as research subjects, the Federal regulations allow such research only when the study concerns: 1) the possible causes, effects and processes of incarceration, and of criminal behavior; 2) prisons as institutional structures or prisoners as incarcerated persons; 3) conditions that particularly affect prisoners as a class; 4) practices that have the intent and reasonable probability of improving the health or well-being of the subject. All such studies must present no more than minimal risk and inconvenience to the subjects, must be reviewed by an Institutional Review Board (IRB) and may often proceed only after consultation with appropriate experts, including experts in penology medicine and ethics and only after the Secretary of DHHS has published his or her intent to approve such research.

[8]In re Gault, 387 US 1 (1967).

[9]*See Bellotti* and *Planned Parenthood, supra* note 4.

[10]Limitations on the rights of parents to refuse care and treatment on behalf of their children are represented by:

Custody of a minor, 379 N.E. 2d 1053 (MA 1978), reviewed and affirmed, 393 N.E. 2d 836 (MA 1979); *but see* In re Hofbauer, 47 NY 2d 648 (1979).

[11]Wyatt v. Stickney, 325 F. Supp. 781 (M.D. Ala. 1971); Donaldson v. O'Connor, 493 F. 2d 507 (5th Cir. 1974), vacated and remanded, 422 US 563 (1975).

[12]*See Rennie* and *Rogers, supra,* note 4.

[13]NY State Association for Retarded Children, Inc. v. Carey, 393 F. Supp. 715 (1975).

[14]Protection of Human Subjects, 45 C.F.R. Section 46 (1983, as revised).

[15]Informed Consent is mandated by common law and Federal regulations: a) 45 C.F.R. 46.116; b) A. Meisel, What It would Mean to be Competent Enough to Consent to or Refuse Participation in Research: A Legal Overview, Presented at the National Institute of Mental Health Workshop (Jan. 12, 1981) at 9; c) President's Commission for the Study of Ethical Problems in Medicine and Biomedical and Behavioral Research, Making Health Care Decisions: Volume I: Report (Washington, DC: US Gov't. Printing Office, 1982).

[16]The elements required for informed consent are found in 45 C.F.R. 46.116.

[17]There is indeed a ready analogue for modifying the requirements of informed consent with patients of questionable capacity in The National Commission for the Protection of Human Subjects of Biomedical and Behavioral Research, Report and Recommendations for Research Involving Those Institutionalized as Mentally Infirm (DHEW Pub. No. 05, 78–0006, 1978). That report concluded that there is a "paucity of knowledge"

relating to the care and treatment of persons institutionalized as mentally infirm and that "clearly, improvements are in order; and these improvements are strongly dependent upon research." (*Ibid.*, at 113). The Commission considering research on the institutionalized mentally infirm, however, was premised on the possible, sporadic, and variable competence of potential subjects. [*See,* US Department of Health, Education and Welfare, Proposed Regulations on Research Involving Those Institutionalized as Mentally Disabled, 43 Fed. Reg. 53950 *passim* (1978)].

The Commission concluded, after considerable debate, that individuals institutionalized as mentally infirm should be able to participate in research presenting more than minimal risk and no direct benefit to them under very limited conditions; only a minor increment of risk (over minimal) may be presented, and the anticipated knowledge to be gained from the research must be of vital importance for the understanding of the condition for which the subjects have been institutionalized or must be expected to provide some benefit for the subjects in the future. In addition, appropriate conditions for the consent or assent of the subjects must be met (including supervision of the process by a consent auditor) and no subject may be included in such research over his or her objection. *Ibid.*, at 120. (Note, however, these regulations were never passed.) For a critique of these regulations, which may account in some measure for their rejection, see "Public Response Critical of HEW Regulations on Mentally Disabled, IRB, April, 1979, p. 8.

[18]*See* Meisel, *supra* note 15.

[19]43 Fed. Reg. 53950 *passim* (1978).

[20]*See* Meisel, *supra* note 15.

[21]45 C.F.R. 46.110 (1983, as revised).

[22]*See* Head v. Colloton, 331 N.W. 2d 870 (Iowa 1983).

[23]Kaimowitz v. Department of Mental Health for State of Michigan, Civil Action No. 73–19434—AW, Circuit Court, Wayne County, Michigan (July 10, 1973), as reported in *Wadlington*, p. 973, 1981.

[24]*Ibid.*, p. 981.

[25]*Ibid.*, p. 984.

[26]*Ibid.*, p. 983.

[27]Skilled Nursing Facilities (SNF), 42 USC 1395 x (j); Intermediate Care Facilities (ICF), 42 USC 1396 d (c).

[28]42 USC 1395 aa(a) and 1396 a(33)(b).

[29]*See* Bruce C. Vladeck, *Unloving Care: The Nursing Home Tragedy* (New York: Basic Books, 1980), for a most comprehensive expose, review of the issues, and analysis of social policy.

[30]David Rothman, *The Discovery of the Asylum: Social Order and Disorder in the New Republic 38* (Boston: Little, Brown and Company, 1971).

[31]W Gaylin, I. Glasser, S. Marcus, and D. Rothman, *Doing Good: The Limits of Benevolence 109* (New York: Pantheon Books, 1971). See also, I. Glasser, Prisoners of Benevolence: Power v. Liberty in the Welfare State, in Doing Good: The Limits of Benevolence 97 (1978).

[32]45 Fed. Reg. 47383 (July 14, 1980), stating: "The facility must protect and promote each patient's right to a dignified existence, self-determination, communication with and access to persons and services inside and outside the facility, and to exercise his or her legal rights."

[33]*See* note 4, *supra*.

[34]Rhoden, *The Right to Refuse Psychotropic Drugs,* 15 Harvard Civil Rights—Civil Liberties Law Review 363, 384 (1980).

[35]Rennie v. Klein, 476 F. Supp. 1294, 1297 (1979).

[36]Robertson, *Organ Donations by Incompetents and the Substituted Judgment Doctrine,* 76 Columbia Law Review 48, 64 (1976).

[37]*Ibid.*, at 65.

[38]Capron, *The Authority of Others to Decide About Biomedical Interventions with Incompetents,* in Who Speaks for the Child: The Problems of Proxy Consent (New York: Plenum Press, 1982), p. 120.

[39]*Ibid.*, p. 136.

[40]*Ibid.*, pp. 133–134.

[41]In re Quinlan, 70 NJ 10 (1976); Superintendent of Belchertown v. Saikewicz, 370 N.E. 2d 417 (Mass. Sup. Jud. Ct., 1977); in the Matter of Shirley Dinnerstein, 380 N.E. 2d 134 (Mass. App. 1978).

[42]*Dinnerstein,* p. 139.

[43]*Ibid.*, p. 138.

[44]In the Matter of Spring, 399 N.E. 2d 493 (Mass. App. 1979).

[45]In the Matter of Spring, 405 N.E. 2d 115 (Mass. Sup. Jud. Ct. 1980).

[46]*Ibid.*, p. 122.

[47]In the Matter of Storar, in the Matter of Eichner, 420 N.E. 2d 64 (NY Ct. of App. 1981).

[48]*Ibid.*, p. 72.

[49]*Ibid.*, p. 72.

[50]*Ibid.*, p. 73.

[51]Any discussion of the appropriateness of a guardian should take the following into account:

1. To be helpless is not necessarily to be in jeopardy. To be helpless and unloved is the matrix of disaster, *DOING GOOD,* p. 19.

2. Small appealing children tend to be most often cared for under guardianship authority, and may easily be distinguished from drooling, incontinent, smelly, and decrepit demented patients. There is also the economic factor of intergenerational conflict and strife. Whereas, children are most clearly a drain on the assets and resources of their parents, they are much less dramatically so than an old senile institutionalized parent, eating away at the family assets at approximately $26,000 per year. The possible conflicts between a guardian–relative and ward are simply more evident and inevitable with the senile and the elderly. "The parent of a healthy six month old child simply does not bear the same relationship to that child that an estranged son-in-law does to the senile, wealthy old lady to whom he is the next of kin." *DOING GOOD,* p. 28.

3. In the context of guardians empowered to refuse treatment for the institutionalized mentally ill, the guardianship process has been found inadequate because, among other reasons:

"It often does not actually protect the patient's right to refuse medication because of the difficulties in finding a guardian who understands the complex issues involved in the decision and who is sufficiently free of psychological, psychosocial or socioeconomic concerns that he or she can act unambiguously in the patient's best interest." Guntheil, Shapiro and St. Clair," Legal Guardianship in Drug Refusal: An Illusory Solution," American Journal of Psychiatry 137 (3), 1980, 347, 350.

[52]Parham v. J. R., 442 US 584 (1979) dealt with the issue of the review required before parents could place their children "voluntarily" in mental institutions. One of its solutions was to provide that "some kind of inquiry should be made by a 'neutral factfinder' to determine whether statutory requirements for admission are satisfied," p. 606. Thus, some "neutral factfinder" could review and determine the adequacy of the facility.

[53]*See* E. Kubler-Ross, On Death and Dying (New York: MacMillan, 1969); B. Soukes, The Deepening Shade: Psychological Aspects of Life-Threatening Illness (Pittsburgh: University of Pittsburgh Press, 1982).

[54]Imbus and Zawacki, "Autonomy for Burned Patients When Survival is Unprecedented," New England Journal of Medicine 197 (6), 1977, 308.

[55]House Bill No. 4058, introduced Feb. 1, 1979:

"A bill to confirm the right to accept or refuse medical treatment; to provide for the appointment of agents and prescribe their powers and duties; to prescribe certain criminal and civil liability; and to provide for certain immunities."

[56]The Natural Death Act, CA Health and Safety Code, Section 7185–7195 (West Supp. 1983).

[57]*Ibid.*, at Section 7188.

[58]*Ibid.*, at 7188.

[59]Winslade, "Thoughts on Technology and Death: An Appraisal of California's Natural Death Act, DePaul Law Review 26, 1977, 717.

[60]Additional arguments in regard to overregulation of medical care, less relevant to research, have also been raised. Consider: "Finally the Committee considered the usefulness of *any* legislation we might propose. In such consideration, we were mindful of the ill effects on society, and particularly on medicine, of over-legislation and regulation or what we might term "legal pollution." We are aware of the often valid complaints of the medical profession that excessive regulation is not consonant with good practice of medicine and can be detrimental to the best interests of the patient."

Committee Report, "Death with Dignity," by the Committee on Medicine and Law, the Bar Association of New York (Feb. 15, 1977).

[61]Gassel, Levy-Warren and Weiss, "Representing the Helpless: Toward an Ethical Guide for the Perplexed Attorney, Western State University Law Review 5(2) 1978, 173, 174.

[62]*Ibid.*, p. 185.

[63]Indeed, there may be an argument that nursing homes, as "public institutions" and the officials in local governments which fund them, and the individuals in government and administration, may be liable under an application of 42 USC 1983, which provides that every "person" who, under color of any statute, ordinance, regulation, custom or usage, of any State or Territory, subjects, or causes to be subjected, any citizen to the deprivation of any rights, privileges or immunities secured by the Constitution and laws, shall be civilly liable to the party injured.

Two arguments support this statement. First, "Nursing homes are not directly run by the government. However, that does not mean that they are any less public institutions . . . the notion that nursing homes are private institutions entering into private contracts with individual patients is a convenient fiction. In fact, nursing homes are the instrument through which the government discharges its obligation to take care of the elderly . . . and . . . should therefore be viewed . . . as essentially another government institution." *Glasser*, note 32 *supra*, at 110.

In addition, consider the decision in Monell v. Department of Social Services of the City of New York, 436 US 658 (1978), extending possible liability to local governments for unconstitutional execution or implementation of policy. Research protocols which invade constitutionally protected rights may thus result in extended liability.

Issues of Equity in the Selection of Subjects for Experimental Research on Senile Dementia of the Alzheimer's Type

Harry R. Moody

Suppose that an innocent man is arrested for a crime he did not commit. He is arraigned before a judge, given full knowledge of his legal rights and provided with a high-quality attorney, who defends his client before an impartial jury of his peers. Nonetheless, the accused ends up being convicted and sent to prison for many years.

Are we prepared to see the outcome of this process as "just?" No, we are not. Despite the fact that fair and just legal procedures have been scrupulously followed, the outcome is a violation of our sense of equity. This example makes clear how we distinguish, in ordinary moral reasoning, between procedural justice—following rules of fundamental fairness—and substantive justice, or what we may simply call "equity."

The problem simply is this: what would count as "equity" in the selection of patients to participate in research on Alzheimer's disease? Is it possible that we could generate rules of procedural justice for carrying out Alzheimer's research and still fail to achieve "equity" or substantive justice in the way in which experimental subjects are selected? Let me bring this question to the context of actual Alzheimer's research with a hypothetical example and case study.

Dr. William Artz is faced with a problem. For the last three years he has been doing research on the neurochemistry of Alzheimer's disease. Preliminary studies have been encouraging, but now there's a need to gather data from a population whose brain function and behavior can be monitored closely over a period of time.

Dr. Artz's research lab in the Medical School has recently entered into affiliation agreements with Goldengrove County Nursing Facility and also with Holy Spirit Home for the Aged. Goldengrove County Facility is a public long-term care institution serving a very large, primarily minority and poor clientele. Holy Spirit Home, by contrast, is a highly regarded voluntary facility with a long waiting list.

Dr. Artz approached the administration of both facilities to determine whether he could gather data from spinal fluid samples from residents in the facilities. Dr. Artz assured both nursing homes that the research team intended to obtain full informed consent from residents willing to participate in the program.

Goldengrove County Facility agreed immediately to participate and seemed eager for more contact with Dr. Artz's Medical School. Goldengrove's Administrator assures Dr. Artz that "his" residents will be happy to participate, especially in view of Dr. Artz's promise of follow-up monitoring and early diagnosis of neurological problems.

The Administrator of Holy Spirit Home, by contrast, was unhappy with Dr. Artz's proposal. After consulting with his attorney and the Board of Trustees, he informed Dr. Artz that the Home could not participate under any circumstances. Members of the Board felt strongly that "their residents wouldn't be used as guinea pigs."

What should Dr. Artz do now? Should he go ahead and administer informed consent protocols to gather data at Goldengrove Facility alone or should he pursue some other course of action?

We may begin by noting the strongly paternalistic stance on the part of both facilities in this case. In the Holy Spirit Home, residents are not even given the opportunity to hear about the experiment and decide for themselves. In Goldengrove the administrator is confident that "his" residents will agree to participate, as perhaps they actually will, in order to get some free, high-quality medical care that might otherwise be unavailable to them.

The paternalism issue must be raised here, not because I endorse paternalism with respect to the institutionalized elderly, but because it is important to recognize the large potential for paternalistic manipulation that is available to administrators, attending physicians, and experimenters by virtue of their control of the communication process and, more basically, by the dependent position of the institutionalized elderly. If this fundamental inequality and dependency is a fact of life, then we can no longer escape considerations of equity simply by establishing that an experimental subject made a choice to participate in an experiment-free decision based on informed consent. The very decision to present or withhold the choice to participate must itself be governed by some prior judgment about broad considerations of equity. A "free" decision by Bowery bums to sell their blood or to participate in medical experiments does not dispense with our obligation to assess the propriety—that is, the equity—of going forward with the procedure in the first place.

In this case study, for example, the contract between Goldengrove and Holy Spirit is painful. The well-endowed, voluntary facility, protected by a powerful board and attorney, resists having its people used as "guinea pigs"—even if the resistance is irrational and might perhaps deprive residents of some marginal benefits. Goldengrove public facility, with its poor and minority clientele, eagerly offers itself for the research. We cannot help being reminded of the experiments at the Brooklyn Chronic Diseases Hospital where elderly patients were injected with live cancer cells in early medical experiments. Perhaps there will be no comparable dire consequences, but a purely consequentialist perspective seems inadequate. Rule-utilitarianism might suggest that over-representation of poor and minority elderly in public facilities is a bad precedent to set and a bad habit for researchers to develop. Does this mean that Dr. Artz should refuse to do research on a predominantly poor and minority elderly population simply because the more affluent and powerful group at Holy Spirit Home have exercised their right to be irrational (as Dr. Artz thinks of it)? If Dr. Artz refuses to conduct research at Goldengrove, after instituting all proper procedures for informed consent, then he would not only fail to pursue a line of investigation with potential promise for enhancing collective welfare; he would also be guilty of the same paternalism shown—in opposing but parallel guises—by administrators of both Goldengrove and Holy Spirit facilities. The more difficult path here would be to pursue research with low risk but promise of great benefit, while respecting all appropriate safeguards of informed consent, just compensation for experimental subjects, and—most important of all—seeking to raise the consciousness of all parties to the inquiry: research team, administration of the facility, and patients and their families.

This last point suggests that undertaking research—even in

conditions of doubtful equity—has itself the quality of "ethical experimentation" in the sense of pushing back the definitional boundaries of equity itself. In the case of an innocent man condemned by a mechanism of procedural justice, we have no difficulty in knowing where substantive justice (equity) will actually lie. The guilty should be punished and the innocent go free. When rules of procedural justice yield outcomes in violation of substantive justice, then a "corrective rule" of equity—for example, the commercial law provision for overturning some properly executed contracts as "contrary to public policy"—must be provided on grounds of equity itself. The problem here is that it is not so easy to say where substantive justice lies in the case of going forward with Alzheimer's research at Goldengrove facility. Do the criteria for selection in this case fulfill the requirements of equity or do they not? "Ethical experimentation" and "consciousness raising" sound like good slogans, but they leave the question unanswered.

Ethics and Equity: The Limits of Regulation

One of the more dismaying articles in recent years is a paper published by Richard Ratzan titled "The Experiment that Wasn't: A Case Report in Clinical Geriatric Research" The Gerontologist, 21, 1981. In that article Ratzan reports on the failure of a clinical research project to recruit elderly institutionalized subjects for participation in a medical experiment of negligible risk (not in the area of Alzheimer's research). Ratzan, in his recruitment process, scrupulously carried out procedures to insure full autonomy and protection of the elderly subjects who were solicited for the research, including "veto power" for private physicians, rejection of proxy consent, a two-part interview design to validate competence of the subjects, and thorough monitoring and review by the medical director of the facility. The final result of this scrupulous adherance to ethical rules and principles was that the recruitment effort utterly failed: no subjects were finally enlisted from among the 312 potential participants initially considered. Ratzan's "meticulous attempt to provide freedom of choice," he admits, could yield the implication that "unethical recruitment is more effective"—a conclusion that he (rightly) refects. Yet his additional conclusion—that long-term care institutions do not represent ideal or easily workable settings for ethical, non-therapeutic clinical research"—could well become a counsel of despair. Solicitation of Alzheimer's patients whose condition can be carefully monitored over a period of years may simply be unfeasible if institutional settings are excluded.

The problem of equity in the selection of patients for Alzheimer's research raises some fundamental questions about ethical principles and rules in the protection of human subjects. Consider the analogy with Type I and Type II errors in hypothesis testing. In our principles for the protection of human subjects, we are concerned to avoid Type I errors—namely, accepting a hypothesis as true when it actually is not. We are less concerned with making Type II errors—rejecting a hypothesis as false when it is actually true. In other words, we err on the side of caution, and perhaps properly so. Are we really prepared, for example, to endure the dangers of cutting corners that may be entailed by all-out competition among scientists, say, for the Nobel prize awarded for applying positron tomography in discovering a cure for Alzheimer's disease? Memory of questionable ethical practices in the discovery of DNA and in the recent history of biological science in general should certainly give us pause.

And yet, the case study of Dr. Artz at Goldengrove and Holy Spirit Homes, like the "experiment that wasn't" reported by Dr. Ratzan, shows us clearly that something has gone wrong in our approach to the protection of human subjects in general. This is a large issue, but Alzheimer's research raises all the general questions and some additional, quite specific ones that deserve consideration. The cases of Dr. Artz and Dr. Ratzan demonstrate, respectively, the kind of Type I and Type II errors that our system of procedural justice was intended to avoid. It is not enough to dismiss these failings as unavoidable imperfections of any system of procedural justice (e.g., rules for the protection of human subjects). We must raise the prior question of whether the Type I and Type II errors here are not in fact unavoidable features of any system of procedural justice. We must ask, in other words, how the claims of equity in the selection of subjects can find proper expression in stable institutions of justice.

I turn here to a recent paper by Stephen Toulmin, "The Tyranny of Principles" (*Hastings Center Report,* 11(6), December

1981. Though Toulmin does not discuss aging research or Alzheimer's disease, his paper grows out of a context appropriate for our current discussion—namely, his work with the National Commission for the Protection of Human Subjects of Biomedical and Behavioral Research.

Toulmin's argument is that we have overexaggerated the function of rules or principles in both law and ethics. Far from playing an essential part in either law or ethics, he argues, rules have a much more limited place. Over against the procedural justice of rules and their justifying principles, there are claims of substantive justice or equity. In Toulmin's words,

> *Justice has always required both law and equity, while morality has always rested on both fairness and responsiveness. When this essential duality is ignored, the insistence on discussing fundamental issues at the level of unchallengeable 'principles' can generate . . . its own subtle kind of tyranny.*

The consequences of this trend are far from beneficent. In our current systems of law, ethics, and public administration, we increasingly prize rule-governed uniformity at the expense of equity. In social services and in the professions generally, there is a steady erosion of the fiduciary conception of professional responsibility and a growing distrust of any individual discretion—a prescription for bureaucracy and the "litigious society." By making uniformity or equality the test of "fairness," we fail to acknowledge the overriding demands of equity, which means doing justice with discretion, as Toulmin observes, "in the interstices of, and in areas of conflict between our laws, rules, principles and other general formulae." The claims of equity give rise to the "balancing test" criterion for the adjudication of rival principles for a model of arbitration and prudent judgment—the essence of discretion and equity—over against an exclusively adversary system of justice.

Toulmin couples his theoretical argument with reference to the ancient Roman pontifex system of justice, where judgments were rendered by a "wise man" brought in to consider the merits and the equities of a given dispute, without regard for an ongoing system of formal rules. Several points must be made in response to this institutional embodiment for adjudicating claims of equity. First, our present-day pluralistic society certainly does not furnish the pre-existing agreement of custom and consensus on which a pontifex system must rely. Second, superseding claims of uniform equality in favor of unequal "equity" seems to work effectively only on a small scale; for example, within a family circle or in analogous contexts—what Toulmin calls—the "ethics of intimacy" takes precedence over the "ethics of strangers." In our present-day society of strangers, an ethics of estrangement—perhaps contracted behind a reasonably thick "veil of ignorance" (Rawls)—may be our best protection from at least Type I errors, say, in selecting subjects for participating in Alzheimer's research. In the "ethic of strangers" we are not prepared to hand ourselves over to the tender mercies and unlimited discretion of Dr. Artz. Yet generating still more rules to protect the unprotected—say, the Goldengrove residents—simply contributes to the problem of overregulation and overprotection, leading in turn to Type II errors, as in "the experiment that wasn't." The limits of regulation and limits of equity both seem painfully evident. Is there some "balancing test" that might lead us beyond this impasse?

Some Principles for Equity in the Selection of Subjects

Let me briefly summarize my argument to this point. The practice of geriatric research on human subjects strongly suggests that our current systems of procedural justice are perhaps necessary, but not sufficient to guarantee substantive justice. We want to protect the vulnerable from exploitation, but at the same time, we do not want to inadvertently screen out those elderly who are capable of bearing burdens for the common good of their own age-group. (I exclude here the question of bearing burdens for the common good of society as a whole, which presents more complicated questions that I cannot consider.) In other words, we need to strike a better balance in avoiding both Type I and Type II errors of equity in selection. Both errors, I stress, are errors of equity (substantive justice), just as much as it is a violation of equity for the innocent to be punished or for the guilty to go free, even if the violation of justice in the two cases is not precisely equal. Finally, we should try to embody in institutional behavior some method for insuring that considerations of equity are not overlooked as we erect a new bureaucratic structure of Institutional Review Boards and the like. Failing to address considerations of equity in the protection of human subjects erodes the legitimacy of this system of distributive justice altogether, just as much as unjust imprisonment, plea-bargaining, or failure to punish crimes serves to undermine the legitimacy of the criminal justice system. Both distributive and retributive justice require concern for substantive as well as procedural justice, but, if Toulmin's argument is right, then our society is increasingly drifting toward a preoccupation with "proper procedures"—and thus, heightened bureaucracy and regulation—without guaranteeing any improvement of equity in the result.

With this formulation of the problem in mind, let me suggest a series of principles for the assessment of equity in the selection of subjects in Alzheimer's research that might address the concerns I have raised.

Concern for the Least Advantaged

Hans Jonas' "descending scale" for selection of human subjects still retains its plausibility and is reinforced by Drew Christiansen's "first principle" of geriatric ethics—namely, "losses are not to be compounded." The losses of elderly people suffering from Alzheimer's disease are already grievous, both to themselves and to their families. What Rawls calls the "difference principle" suggests that Alzheimer's research be undertaken on this group only when unavoidable and when the promise of benefit is very great. When the research is nontherapeutic (i.e., the benefits flow to others than the participants in the research, even if members of the same representative group, such as future sufferers of Alzheimer's disease), then we ought to provide current benefits—such as additional medical care—to those who volunteer to participate in the research effort. Needless to say, all appropriate procedures to insure informed consent—or where unavoidable, proxy consent—must be followed to respect the autonomy of those who participate in the research. These issues of consent are discussed by others in this volume and I will not address them here. Questions of consent aside, however, I am proposing that every experimental intervention be judged as to whether it improves the well-being of the least advantaged members of the Alzheimer's research population

(Rawls' "difference principle"). On this criterion, we might well approve Dr. Artz to go forward with low-risk experiments at Goldengrove facility.

Compensation

The general issue of compensation, especially from a legal point of view, is treated by others in this volume. One aspect of "just" compensation, I argue, reflects our obligation to improve the position of the least advantaged, but the compensation problem includes other dimensions of equity as well. Compensation of research subjects should not be seen exclusively as a matter of payment for incurring risks or as a guarantee of restitution in cases of harm, negligence, or otherwise. I call both these elements of the compensation principle ("fair" compensation) the insurance dimension of compensation, and they both constitute limits of equity that we ought to apply to any allegedly "free" decision of an experimental subject (or his or her proxy) to participate in an experiment.

But the insurance principle is not enough. The demand for equitable compensation must take account of the de facto limits on the free choice of individuals to undertake risks or bear burdens either for their own benefit, or for the common good. This demand need not result in paternalistic second-guessing of informed consent, but it requires that we look systematically at the collective outcomes of those individually free decisions. If a disproportionately high number of poor ghetto blacks find themselves plea-bargaining or receiving longer jail sentences, it must make us wonder about the "free" consent of the accused, though it need not make us forbid individuals to plead guilty to crimes. The demand for equity in compensation, again, requires that we look at outcomes, not simply proper procedures, and that we look at collective consequences, not simply individual costs, benefits, or choices.

The stress on the collective significance of compensation opens up another dimension that I alluded to in the matter of bearing burdens for the common good. The city of Miami now provides free transportation for witnesses or crime victims on vacation in Miami to come back to the city at a later date in order to participate in the trial of those accused of the crimes. Charitable organizations have for some time provided free transportation to blood banks or points where individuals can make blood donations. Both trial testimony and donating blood are modest examples of how society provides a kind of compensation in order to facilitate voluntary contributions to the common good. A principle of this kind should perhaps be applied to Alzheimer's research selection in order to cast a wider net in recruiting subjects for experimentation.

For too long our ethical and political discourse has been dominated by the first two terms in the slogan "Liberty, equality, and fraternity." Liberty (or autonomy), equality of treatment, and opportunity are certainly indispensable; they constitute the first two principles that Rawls ascribes to this theory of justice. Yet liberty—in the sense of freedom from restraints—and equality—as egalitarian treatment—are frequently in conflict. On the current scene, liberty demands a reduction of rules and bureaucracy, whereas equality seems to call forth more rules and bureaucracy.

To insist on procedural safeguards for individual informed consent is to exalt liberty (freedom) as the ultimate value. To demand equality of treatment for all potential experimental subjects is to exalt egalitarianism as the ultimate value. Both principles are essential. Yet we have lost sight of the third ideal—fraternity, or the sense of participating in a common enterprise that calls forth sacrifices on behalf of the common good. For geriatric experimentation in general, this last principle becomes crucial, for the limited life expectancy of the elderly means that it will normally be the case that individuals who participate in experiments will not benefit much from the new knowledge gained, even if future generations of elderly are benefitted.

The evolution of liberalism has favored the first two principles—liberty and equality—at the expense of the third. Procedural due process guarantees—such as informed consent—are intended to assure maximum liberty, whereas the egalitarian initiatives of the welfare state are to improve the welfare of the least advantaged. Together we are to have a just state. The disillusionment with the ambiguous results of "procedural" liberalism and its penchant for regulation has now reached a point where all forms of regulation are increasingly challenged. In this paper, following Toulmin, I have offered arguments sympathetic to an "antiregulatory" mood, but I am not blind to the dangers concealed down that road. A society based on institutionalized fraternity and adjudications of equity remains elusive. In two additional principles for equity I will try to address this demand for substantive, as well as procedural, justice.

The Voice of the Afflicted

If the principles of liberty (informed consent) and equality (equal protection) are to be broadened to the collective principle of fraternity, then the ideal of self-determination must itself be broadened. Why do we imagine that individual informed consent—to procedures, experimental goals, methods of communication, and so on—devised by younger research scientists will adequately ("equitably") reflect the wishes of those afflicted with Alzheimer's disease?

On the contrary, in place of this case-by-case individualistic notion of consent (liberty), I argue for a collective expression of self-determination through an appropriate institutional channel. The demands of equity require not merely a "yes" or "no" response (like an election) to options (candidates?) pre-selected by an elite of experimenters. Instead, equity requires us to examine the experimental inquiry itself, in its nontechnical dimensions, reflecting the voice of the afflicted—the voice of those who will bear the burdens of this research. Providing a voice for the afflicted applies at the collective, as well as the individual, level. It demands not merely informed consent, but participation in the policy framework in which decisions of consent arise in the first place.

The example of an innocent man sent to prison or a vulnerable population group (Goldengrove) selected for experimentation shows that procedural justice can have imponderable and unpredictable effects. Well-intended schemes for protecting human subjects may have the effect of preventing socially desirable experimentation ("the experiment that wasn't") that could materially improve the position of the least advantaged. Similarly, some versions of "truth-telling" that barrage experimental subjects with so much information that they become frightened or confused obviously violates the spirit, if not the letter, of procedural requirements for informed consent. But who is to determine just how "truth" is best communicated without compromising ethical principles? Too much protection via red tape can delay procedures so that needed research is not done. Too little protection leads to the excesses we are all familiar with (e.g., the Chronic Diseases Hospital case).

The corrective for unintended consequences and the path to democratic participation—the voice of the afflicted—in designing the policy framework is one and the same—namely, systematic review of consequences even in cases where the system seems to be working. The organ for such review of unintended consequences—for assessment of equity and substantive justice—should be a Review Board composed of persons who thoroughly understand the unintended consequences (if any) of our well-intended schemes of research efficiency or legal protection. In this case, I propose that the Review Board be composed of those elderly people who intimately understand the consequences of Alzheimer's disease—not the sufferers themselves (for reasons of incompetency), but an appropriate proxy—elderly people who have seen the consequences of the disease in their known immediate family. Such people, I believe, would be sensitive to the horrors of the disease and eager to promote any advances that might spare others from their plight. At the same time, being elderly themselves and potential victims, as well as potential research subjects, they would have a keen interest in forms of experimentation that will insure equity in practice (not merely by proper procedures). Needless to say, such a Review Board ought to include representatives from special groups—such as residents of publicly-funded nursing homes, minority groups, those without families, and so on—who might be at special risk.

I am less concerned with the juridical status of this Review Board (advisory, court of appeals for alleged ethical violations, ombudsman, and so on) than I am with arguing for its existence and for the importance of such representation in our design and implementation of experimentation on human subjects. There is, admittedly, a tension between the meritocratic decision-making of science and the democratic principles embodied in my "civilian review board." But scientific efficiency, even if coupled with legalistic procedural guarantees of consent and due process, will not provide for concern for the common good, nor for the claims of equity in the selection of those who bear the burdens on behalf of the common good.

Internalizing Norms of Equity

I have argued here repeatedly for acknowledging the dependency and vulnerability of the institutionalized elderly who are probable participants in Alzheimer's research programs. I do not endorse this dependency and vulnerability, but I regard it as a nearly unavoidable human reality in the nonideal context of actual research activities. Ratzan, in his comments on the "subtle forces of paternalism, institutional coercion and dependence," recognizes the same human reality.

A measure of civilization is the protection of the powerless under the rule of law. Nothing of what I have urged here should be construed as a belief that the powerless, e.g., in nursing homes, would be better off without laws, principles, rules, regulations, and other restraints on those who are tempted, even if unwittingly, to exploit their vulnerability. Yet, with Toulmin, I argue, too, that our laws must operate so as to encourage concern for equity, or, what comes to the same thing, encourage a habit of self-restraint and self-examination. When we speak of the fiduciary role of the professional and the function of discretion in that role, we already presume the existence of these habits of self-examination and self-restraint. The point is not to do away with law or external restraints on conduct, but instead to raise the question: How can law (rules, regulations, formal restraints) promote the internalization of translegal values such as equity so that the selection of vulnerable subjects for Alzheimer's research may have some hope of yielding equitable outcomes and not just "proper procedures"?

Arthur Caplan recently made a "modest proposal" for the reform of Institutional Review Boards, which suggested one answer. Instead of requiring prior approval (and burden of proof, paperwork, red-tape, etc.) from investigators, we should move toward a post-audit system for monitoring compliance with standards and principles for the protection of human subjects. Post-audit systems are widely used in many walks of life. The branch manager of a bank, for example, understands perfectly well the standards and principles for protecting the bank's capital, but he is encouraged by the management to make his own decisions in conformity with the policy guidelines previously established. What makes this system work is that the threat of random auditing by bank auditors who can, at a moment's notice, swoop down on a local branch and conduct a full-scale examination of books and records.

The value of all such "post-audit" systems is that they encourage the internalization of norms and standards of behavior of human conduct. The most impressive example of such a "post-audit" system is the organized activity of scientific investigation itself. Cheating and fraud will occasionally occur, but rarely enough, in view of the threat of exposure and humiliation for those who violate the norms of scientific research, as some recent cases in biology serve to remind us. The current system of prior approval, rather than post-audit, examination shifts all the attention of Institutional Review Boards to processing paperwork, leaving negligible time for actual "field audits" to observe the nature of compliance by investigators. Not only is this system a poor use of limited enforcement resources, but it puts all our attention on issues of procedural justice, rather than equity and substantive justice. By stressing prior approval and paperwork, it fails to encourage habits of thought leading to self-examination, self-restraint, refinement of discretion, and the like. We overlook such virtues at our peril.

All of these virtues, however old-fashioned they may sound, are ultimately our only guarantee that compliance will be real and not merely nominal. The practice of scientific research is too decentralized for purely legalistic and procedural remedies to guarantee that the least advantaged will actually be protected if scientific investigators themselves begin to view protection of human subjects as a matter merely of paperwork and red-tape. Ratzan's "implicit deduction"—namely, that unethical recruitment is more effective—should continue to haunt us and disturb the dreams of reason that concoct schemes of social enforcement. Such arrangements depend on deeper schemes of social cooperation—mutual trust, mutual criticism, and equity—that must ultimately govern our lives.

Alasdair MacIntyre, in, *After Virtue* (Notre Dame, 1981), argues along a line not unlike Toulmin. Self-examination, self-restraint, discretion, and judgments of equity ultimately spring from the cultivation of what the philosophic tradition back to Aristotle has known as the virtues. Without the virtues—courage, truthfulness, compassion—the protection of the powerless by the rule of law will not be enough. Without some scheme of social cooperation promoting these virtues, equity in the selection of Alzheimer's subjects may not prove possible in practice, whatever reports and procedures may lead us to imagine.

I conclude with a Chinese proverb: When the Tao disappears, morality remains. When morality disappears, law remains. When law disappears, force remains. When force disappears, chaos covers all. It may be too much to expect that a pontifex system or the rule of the Tao will gain much hearing in our present world. But the balance between law and morality must continually be reexamined. The protection of the powerless depends upon it.

Part 5

Competency to Give Consent

Competency to Consent to Research

Barbara Stanley

Introduction

A determination of competency must be made prior to recruiting an individual as a research subject. If a person is considered to be incompetent, special protection must be afforded him or her before participation in research is permissible. By convention, this special protection is provided by obtaining the consent of a legal guardian, if one has been appointed, or a competent close relative. Thus, some form of competent consent is a prerequisite for almost any research and as a result the determination of competency plays an important role in the research process.

Despite that importance, both the formulation of what constitutes competency and our tools for its assessment remain poorly developed and not well-systematized. Investigators frequently find themselves in the position of having to assess competency with only minimal guidance from the literature and little knowledge of available techniques. As a result, competency assessments are sometimes carried out on the basis of the investigator's personal standards, and it is possible to find one investigator excluding a patient from a research project while another would include that same patient. These difficulties can be overcome, in part, by the use of explicit standards that are closely related to the task at hand rather than of some general assessment of mental functioning. Such a standard can be flexible, depending on the level of risk inherent in the protocol. Furthermore, explicitly stated standards and a standardized means of assessment offer some protection to investigators who are at times questioned by outside monitoring sources for including possibly incompetent subjects in their research protocols. Nearly every time a research protocol involving potentially "vulnerable" subjects is reviewed by the Institutional Review Board (IRB), a question is inevitably and justifiably raised: "Are these subjects capable of giving consent?" Such protocols are approved only with a good deal of uneasiness among IRB members. If each IRB knew that certain quite explicit standards were to be applied when assessing competency, their members would be less reluctant to grant approval.

The focus of this paper then will be to review the standards for assessing competency that have been outlined in the biomedical and legal literature and to suggest techniques for their assessment. In addition, it will identify the possible impairment in competency that senile dementia patients might be found to have, depending on the standard that is employed. However, prior to doing this, one cautionary remark must be made. Although some speculations will be made about possible level of consent capacities in Alzheimer's patients based on degree of impairment, it must be remembered that there is a good deal of individual variability. More importantly, it is crucial to determine empirically whether these speculations are borne out by actual data. Thus, we must study the consent process in Alzheimer's patients with varying degrees of impairment and correlate that impairment with ability to consent. It is particularly important to do so if we are to consider special safeguards for this patient population.

Competency and Consent

Society holds as a goal the preservation of dignity and personal freedom for all individuals, including the elderly and infirm. In the research setting, this is accomplished by permitting individuals to make their own decisions about research participation and to make contributions to research knowledge where they see fit to do so. To facilitate the achievement of this goal, we should not provide protection where it may be unwarranted. Every time an investigator seeks the signature on a consent form of a spouse or child of an elderly research subject, the independence of that individual may be compromised. There are times when this procedure is proper, but it would be prudent to judiciously apply this measure.

Therefore, it can be seen that safeguards without empirical substantiation conflict with one of our societal goals—preservation of individual independence. In this regard, empirical evidence can sometimes present findings that conflict with common notions of patients' abilities in the consent process. An example of this can be seen in the institutionalized mentally disabled who have been identified as a vulnerable population (National Commission, 1978)[1] requiring special protection. However, in a large-scale study of competency, this population was found to make decisions with respect to participation in research that do not differ in any substantial way from medical patients without psychiatric illness. This finding is particularly striking when you consider that the vast majority of the psychiatric patients were actively psychotic.[2–4]

In looking at other empirical evidence that is available with respect to competency to consent in vulnerable populations, we find a somewhat mixed picture. The vulnerable population most researched is the mentally ill. One conclusion at this point, which can be drawn with respect to the mentally ill, is that they do no better than medical patients in the consent process. The evidence that they are less able to give consent is somewhat equivocal and, to a certain extent, depends upon the definition of competency utilized. With respect to the comprehension of consent information, a few studies have assessed psychiatric patients' ability to understand consent information.[5–8] In general, patients do not have a very high level of understanding of consent information. However, when comparing studies of medical patients' comprehension with studies of psychiatric patients, understanding in both groups seem to be fairly equal.[6,7] An example of this is seen in one study that found that schizophrenic patients understood about 50% of the material on a consent form which was read to them.[6] In a direct comparison of psychiatric and medical patients, it was found that schizophrenic patients were more aware of the risks and side effects of their medication than were medical patients.[9] On the other hand, medical patients were better informed about the name and dose of their medication as well as their diagnosis. Related to studies of comprehension are investigations of the literacy of psychiatric patients. Despite the fact that psychiatric patients' comprehension of consent information seems to be equal to medical patients, research indicates that their reading comprehension scores were only at the fifth-grade level.[10,11] As a result, a suggestion has been made that hospital documents be simplified for psychiatric patients,[10] as some have suggested for medical patients.

In studies of psychiatric patients' ability to consent to hospitalization, the results indicate that the level of knowledge of patient rights is relatively poor.[12,13] However, it is important to know whether medical patients would score higher than psychiatric patients and also it is important to separate out what was the result of patients' inabilities from deficient information-giving on the part of the hospital admissions service. In contrast to the studies that conclude that psychiatric patients may not be competent to give consent, one study reports that 93% of the patients gave a valid consent.[14] However, the standard for competency was set much lower than the other studies described here.

In a study of consent to electroconvulsive shock therapy,[8] it was found that about 25% of the patients were incompetent based on their understanding of consent information and independent judges' opinions about their comprehension. This study is the first that has taken a comprehensive approach by coordinating objective information (i.e., patient comprehension) with legal judgments and psychiatric opinions and seems to be a fruitful direction for further research.

A few studies of psychiatric patients have examined the relationship between understanding and the decision to consent or refuse the proposed procedure.[6,8] They found that, like medical patients, psychiatric patients who understood more of the consent information tended to agree to the procedure more often.

With respect to patients' rationale for deciding to agree to a treatment or research protocol, results are not clear-cut. One study found that the risks of psychotropic medication did not play a role in patients' decisions to refuse medication.[5] Psychological factors were cited as primary reasons. It is difficult to compare these results with those from medical patients because no medical study to date has attempted to delineate the psychological factors examined in the psychiatric study. In a study[2–4] that examined psychiatric and medical patients' willingness to participate in a series of hypothetical studies, no differences were found between the two patient groups. Both psychiatric and medical patients agreed to participate in the studies in a manner that was consistent with the level of risk attendant to the study protocol. It is important to conduct a parallel study that investigates participation rate in actual projects.

Overall, the empirical research on competency shows that psychiatric patients do have some limitations on their ability to consent. However, these limitations do not seem to be much different from those found in medical patients.

Similar studies in the geriatric population and with dementia patients need to be conducted. It is also possible that, in the other than severely impaired, we may also find that consent capacities are preserved to a large degree. Efforts are being made to assess ability to consent in the normal elderly.[15,16]

In addition, a study was conducted contrasting nonpsychiatric elderly medical patients' competency to consent to research with that of younger medical patients. This research found that the elderly comprehended significantly less consent information, although their actual decisions about which studies they preferred to participate in did not differ from those chosen by the young. In addition, in a preliminary study of consent capacities of Alzheimer's patients, it was found that patients with moderate impaired have a compromised capacity to consent.[18]

Tests of Competency

Several standards of competency to consent to treatment have arisen from the legal and, to a lesser extent, medical literature. Recently these standards have been adapted to the research setting. The tests are differentiated by the level of protection they provide to the patient. The protection inherent in each test is inversely related to the degree of independence granted to the research subject. In reviewing the standards, I will discuss the level of protection that each test affords.

Five basic standards of competency have been proposed: (1) evidencing a choice; (2) factual comprehension; (3) rational reasoning and manipulation of information; (4) appreciation of the nature of the situation; and (5) reasonable outcome of choice. Except for the last test, reasonable outcome, these tests are ordered here according to the level of protection afforded, ranging from the lowest to highest level. The reasonable outcome test does not fit comfortably at any one point in the hierarchy for reasons that will be explained later.

Evidencing a Choice

This test focuses only on the presence or absence of a decision. If a person makes a decision, he or she is judged to be competent. This test is the least protective of the individual because assessment is minimal.

In actuality, this test is rarely used. It has been criticized on the grounds that this decision does not assure that the patient has a good understanding of the proposed treatment or experiment.

In relation to senile dementia patients, only the most severely incapacitated would be found to be incompetent by this standard. Only those patients who can barely speak would fall into this category. Almost all others would be judged as competent.

In testing competency according to this standard, as mentioned earlier, the assessment effort is minimal. The relevant information is simply presented to the prospective participant, who either agrees or declines to become a subject. No effort is made to determine whether the information was understood or whether rational reasoning was employed to reach the decision. The "Yes" or "No" response by the individual is taken as sufficient to determine competency. Despite some arguments in favor of this test, it seems that, as a society, we feel this standard is too lenient and so we tend to demand a higher degree of competency by prospective subjects.

Factual Comprehension

This test requires that the patient understand the information relevant to the proposed treatment.[19,20] By that, it is meant that the person understands the procedures, risks, benefits, and alternatives available. This test does not take into account how reasonable the decision is or how rational the thought processes were in arriving at the decision. As a result, it is fairly respectful of the research subjects' sense of independence. This test has been the most commonly used standard and is frequently thought to be the only way of assessing competency. In fact, almost all empirical studies of competency and informed consent employ this standard.[21,22]

Despite its widespread acceptance, a major difficulty exists with the comprehension standard of competency. It relies very heavily on an individual's verbal skills, and particularly on verbal expression. Though verbal expression may be a desirable capacity, it is not what the comprehension standard is designed to test. It should test the individual's understanding of information, not his or her ability to convey that information. The confusion of comprehension with ability to express oneself may account for research findings that have shown that ability to demonstrate comprehension of consent information increases with the intelligence of the subjects.[4] As a result, there is a potential for biasing of competency findings against the less verbally skilled. The policy implications are striking—if a person does not have an acceptable IQ, he or she would be likely to be judged incompetent, and thereby be excluded from research, perhaps needlessly. Comprehension can be subdivided into: understanding of the consent information at the moment of consent and retention of that information over a period of time.[23] Retention, however, has been highly criticized as a valid criterion.[24] Although it is important that a research subject remember that he or she can withdraw from an experiment, it may be completely unnecessary that the subject keep all the consent information in mind several weeks or months after the initial decision. The fact that an individual forgets information following a decision does not mean that it was not used during decision-making. It may be more indicative of the normal forgetting process.

In applying the comprehension test in the research setting, the assessment of competency is made by asking the prospective subject to demonstrate an understanding of the consent information. This can be accomplished in a variety of ways that are here presented from least to most difficult: (1) the patient can be asked to repeat the information that was given to him or her while an information sheet or consent form was available; (2) the patient can be asked to do the same without any material available or; (3) the patient can be asked to demonstrate that he or she can utilize the information to draw inferences or conclusions.[25] Procedures for accomplishing this range from an informal interview to teaching sessions with multiple choice tests. With respect to Alzheimer's patients, it is likely that the mildly impaired would be judged competent as long as retention of information was not a criterion. Because memory loss is one of the most prominent clinical symptoms in Alzheimer's disease, both immediate and long-term recall would be impaired. Those patients with a significant degree of impairment would be likely to be judged incompetent because of this. Further, Alzheimer's patients, with their memory problems, present a significant problem if retention of information is a criterion. Do they remember over the course of a study that they are in fact in a study? It seems that information sheets for the subjects to retain that describe the research protocol may be vital for this patient population. In addition, a simplification of the consent information, as recommended in the Preamble of the DHHS regulations[26] may make the information more understandable to the subjects and result in increased competency of the subjects using this standard.

Rational Reasoning and Manipulation of Information

Under this standard, competency is defined as the capacity to understand the nature of the procedure, to weigh the risks and benefits, and to reach a decision for rational reasons.[19,20] This test focuses on the overall pattern of thought rather than the particular result of the decision. It is paternalistic in nature and its application may express a bias toward a particular type of reasoning. In addition, it is difficult to distinguish rational from irrational reasons, verbalized reasons from underlying, unstated reasons, and real reasons from false reasons.[19] Different strategies for assessment of rationality have been proposed.

Although some authors have suggested that the rationality of the individual's entire reasoning processes be assessed under this standard,[27] others have recommended that assessment be tied more directly to the decision to participate in the research.[20,24] The former would be tested with a typical mental status examination designed to elicit severe impairment in judgment, thought disorder, hallucinations, and delusions. Actual judgment in daily life activities may also be assessed. The major problem with this sort of generalized testing is that we do not know, when judgment is impaired in relation to some decisions, whether it is also impaired with respect to decisions in an unrelated area of functioning. In fact, we have a good deal of evidence to the contrary. This brings us to the second form of "rationality" assessment—rationality with respect to the decision to participate in research. In this instance the patient's reasoning about the decision to participate in the protocol is directly elicited. The investigator must assess whether the patient's reasons make sense. A patient who clearly understood the risks and benefits of a protocol and said he decided to participate because he thought the investigator was a "Savior sent from heaven" would be judged incompetent. As an aside, this same patient would be assessed as competent according to the "comprehension" standard as well as the "evidencing a choice" test. Although the example given here is extreme, a difficulty arises when the case is less clear-cut. The determination of what is rational and what is not is left in the hands of the competency reviewer. It has been suggested that the assessor's view of what is rational becomes the standard rather than some objective determinant of rationality. Alzheimer's patients who demonstrate more than a moderate degree of impairment might have trouble reaching this threshold, particularly if a generalized assessment is performed. Since as the disease progresses judgment capacities diminish, reasoning with respect to research participation may become less logical.

Appreciation of the Nature of the Situation

This standard of competency to give consent is closely aligned to the type of assessment that is performed when determining whether an individual is responsible for a criminal act. When applied to competency to give consent, the subject is tested to determine whether he or she appreciates the situation.[27] Does he understand the consequences of consenting or not consenting? Does he understand what information is relevant to his decision and what is not? Under this standard, comprehension of the consent information is a prerequisite with the additional requirement being that the patient must be able to utilize that information in a rational manner. This test is highly protectionistic in that the demands placed on the subject are stringent. For example, it is not sufficient that a subject understand the risks, benefits, and alternatives to the research, the subject must also "appreciate" that he or she is a research subject and what that implies. It has been shown that even "nonvulnerable" populations have a difficult time appreciating that fact.[28]

In applying this standard of competency, the investigator must first establish that the prospective subject has an adequate understanding of the consent information and would then go on to assess the degree to which the patient appreciates that he or she is a research subject, has a certain condition that is being investigated, and knows what will happen if participation is decided upon or decline.

Patients diagnosed as having Alzheimer's disease may be likely to fail this test if they are moderately or severely impaired. Their ability to perform the type of abstract reasoning demanded by this test is likely to be impaired. However, patients with mild impairment seem likely to be judged competent.

Reasonable Outcome of Choice

This test evaluates the patient's capacity to reach the "reasonable" result. The person who fails to make a decision that is roughly congruent with the decision that a "reasonable" person would have made is viewed as incompetent.[20] This standard is applied quite frequently when nonexperimental treatments are recommended and particularly when the patient's life is at stake. It has been highly criticized for its strong paternalistic orientation and, hence, lack of respect for the individual's rights.[20] Any decision with which the competency reviewer disagreed might provide a basis for judging the individual incompetent. Its relevance to competency to consent to research can be problematic since research projects are, by their very nature, experimental. Thus, it is difficult to know what is "reasonable."

However, there is a clear advantage to this test in that it does not rely on the patient's skills at verbal expression as does the comprehension standard, while at the same time it provides more of an assessment than the evidencing a choice criterion. Thus, the patient simply gives a Yes–No response with some prior assessment of whether that is a reasonable response. It has recently been suggested that this standard could be adapted in such a way as to tap "capacity to reach reasonable decisions" about research participation, independent of the actual decision at hand.[4] A patient's ability to determine what are reasonable research protocols in which to participate is assessed. If the patient can make reasonable choices based on a standardized criterion, then that patient is judged to be competent and whatever decision is made about the protocol for which he or she is being recruited is respected. This proposal is novel and differs from the current manner of assessment in which the investigator or competency reviewer decides whether the patient's decision about participation in the research protocol is reasonable. Current practice under this standard would require that the reviewer make some form of assessment of the potential risks to the individual, taking into account the patient's present condition, physical illnesses, and emotional state. Assessment in this format is rarely used because of the high degree of subjectivity on the reviewer's part and the strong paternalistic orientation. However, when one contrasts the current practice with the novel application of this standard, the paternalism inherent in the two applications is quite different. That is the reason this test does not fit into our hierarchy of competency tests. Although the current practice yields a highly paternalistic test, the newer approach is much more respectful of the individuals' sense of autonomy. If individuals can demonstrate that they have the capacity to make reasonable decisions, then it is left in their hands to decide what choice should be made about the present research, what elements they should consider, and what reasoning they should employ. It seems that Alzheimer's patients would be likely to be judged competent up to and including the moderate range of impairment. Memory would not be a major factor in this standard, whereas practical problem-solving capabilities would be assessed.

Conclusion

This paper highlights the major tests of competency to consent to research. Possible impairment in the consent capacities of Alzheimer's patients was noted.

Two prominent areas which ought to be considered in addressing informed consent and competency in Alzheimer's disease:

1. Assessments of competency should be performed in a consistent and standardized manner. The standard used to assess competency should be explicitly identified.
2. Empirical evidence exploring the capacity of Alzheimer's patients to give consent should be collected. It is important to objectively assess what impairments these individuals have in order to accurately determine relevant competency standards.

Note: This work was supported in part by N.I.M.H. grants MH37182 and MH37983.

References

[1]National Commission. *Research Involving those Institutionalized as Mentally Disabled.* DHEW Publ. No. OS 78-0006, 1978.

[2]B. Stanley, M. Stanley, N. Schwartz, A. Lautin, and J. Kane, "The ability of the mentally ill to evaluate research risks." *IRCS Medical Science: Clin. Pharmacol. Ther.; Psychol. Psychiat.; Surgery Transplantation,* 8, 1980, 657–658.

[3]B. Stanley, M. Stanley, A. Lautin, J. Kane, and N. Schwartz, "Preliminary findings in psychiatric patients as research participants: A population at risk?" *Am. J. Psychiat.* 138 (5), 1981, 669–671.

[4]B. Stanley, M. Stanley, C. Peselow, J. Kane, and J. Stein, Informed consent and psychiatric patients: Empirical evidence. *Arch. Gen. Psychiat.* in press.

[5]P. Applebaum, and T. Gutheil, "Drug refusal: A study of psychiatric inpatients." *Am. J. Psychiat.* 137 (3), 1980, 340–346.

[6]L. Grossman, and F. Summers, "A study of the capacity of schizophrenic patients to give informed consent." *Hosp. Commun. Psychiat.* 31 (3), 1980, 205–207.

[7]D. A. Soskis, and R. L. Jaffe, "Communicating with patients about antipsychotic drugs." *Compr. Psychiat.* 20, 1979, 126–131.

[8]L. Roth, C. Lidz, P. Soloff, and K. Kaufman, "Competency to consent to and refuse ECT: Some empirical data. Paper presented at the Annual Meeting of the Am. Psychiat. Assn., 1980.

[9]D. A. Soskis, "Schizophrenic and medical inpatients as informed drug consumers." *Arch. Gen. Psychiat.* 35, 1978, 645–647.

[10]A. Berg, and K. Hammitt, "Assessing the psychiatric patient's ability to meet the literacy demands of hospitalization." *Hosp. Comm. Psychiat.* 31 (4), 1980, 266–268.

[11]G. Coles, L. Roth, and I. Pollack, "Literacy skills of long-term hospitalized mental patients." *Hosp. Commun. Psychiat.* 29, 1978, 512–516.

[12]P. Applebaum, S. Mirken, and A. Bateman, "Competency to consent to psychiatric hospitalization: An empirical assessment." *Am. J. Psychiat.* 138, 1981, 1170–1176.

[13]A. Palmer and J. Wohl, "Voluntary admission forms: Does the patient know what he's signing?" *Hosp. Commun. Psychiat.* 23, 1972, 250–252.

[14]S. Dabrowski, K. Gerard, S. Walczak, B. Woronowicz, and T. Zakowska-Dabrowska, "Inability of patients to give valid consent to psychiatric hospitalization." *Int. J. Law Psychiat.* 1 (4), 1978, 437–443.

[15]H. A. Taub, "Informed consent, memory and age. *The Gerontologist.* 20, 1980, 686–690.

[16]H. A. Taub, G. Kline, and M. Baker, "The elderly and informed consent: Effects of vocabulary level and corrected feedback." *Exp. Aging Res.* 7, 1981, 137–146.

[17]B. Stanley, J. Guido, M. Stanley, and D. Shortell, The elderly patient and informal consent: Empirical findings. *JAMA,* 252, 1984, 1302–1306.

[18]B. Stanley, M. Stanley, and N. Pomara, Informed consent and elderly phy-psychiatric patients. Paper presented at the American Psychiatric Association Meeting, Los Angeles, May 1984.

[19]L. H. Roth, A. Meisel, and C. W. Lidz, "Tests of competency to consent to treatment," *Am. J. Psychiat.* 134 (3), 1977, 279–284.

[20]P. Friedman, "Legal regulation of applied behavior analysis in mental institutions & prisons." *Arizona Law Rev.* 17, 1975, 39–104.

[21]J. Bergler, C. Pennington, M. Metcalfe, and E. Fries, "Informed consent: How much does the patient understand?" *Clin. Pharmacol. Ther.*, 27, 1980, 435–439.

[22]B. R. Cassileth, R. B. Zupkis, K. Sutton-Smith, and V. March, "Informed consent—why are its goals imperfectly realized?" *New England J. Med.* 302 (16), 1980, 896–900.

[23]G. Robinson and A. Merav, "Informed consent: recall by patients tested post-operatively." *Ann. Thoracic Surg.* 22, 1976, 209.

[24]B. Stanley and M. Stanley, "Psychiatric patients as research participants; protecting their autonomy." *Comp. Psychiat.* 22 (4), 1981, 420–427.

[25]G. D. Mellinger, C. L. Huffine, and M. B. Balter, "Assessing comprehension in a survey of public reactions to complex issues." *Institute for Research in Social Behavior,* 1980.

[26]Department of Health & Human Services. *Final Regulations Amending Basic HHS Policy for the Protection of Human Research Subjects.* Federal Register 4616, 1981, 8366.

[27]P. Applebaum and L. Roth, "Psychiatric issues in competency & informed consent. Paper presented at NIMH conference "Empirical Research on Informed Consent with Subjects of Uncertain Competence" January 12, 1981.

[28]A. McCullum and A. Schwartz, "Pediatric research hospitalization: Its meaning to parents." *Pediat. Res.* 3, 1969, 199–204.

Assuring Adequate Consent

Special Considerations in Patients of Uncertain Competence

Alan Meisel

Disease, pursuit and application of knowledge, ethics, and law—these are the four component parts of the problem that this conference is attempting to address. My own efforts are directed primarily at the legal component, and more specifically, at one of its subcomponents.

Outline of the Legal Structure

The subject of this article is the problem of performing research on persons of uncertain competence. This problem is intimately connected with, and an aspect of, informed consent, which is itself part of the larger processes of the legal regulation of biomedical research and medical decision making. Thus, before addressing the problem at hand, it will help to briefly explain the larger context of which the problem of competence is a component part.

The Two-Tiered Approach to the Regulation of Medical Research

The threshold legal problem is whether or not medical research in general is even permissible. Although the older common law took a rather restrictive view of the matter by imposing liability on doctors for departing from accepted methods of medical practice,[1] the contemporary federal statutory law—which, though not formally preempting state legal regulation of medical research, largely governs its conduct today—not only permits, but encourages, medical research. When research is federally funded, its conduct is governed by a fairly specific set of regulations.[2]

The conceptual approach of these regulations is two-tiered. First, the regulations require that a determination be made whether or not a particular research project ought to be permitted to be undertaken by those who propose to carry it out. Implicitly, the following question is asked: "Is this the kind of research in which any individual ought to be permitted to participate?" In answering this question, the decision maker [that is the Institutional Review Board (IRB) as it has come to be known] is directed to focus its attention on a balance of the risks potentially posed by the research and the benefits to individual subjects and/or to society as a whole to be gleaned from the research.[3] If the answer to this question is in the negative, the research may not be conducted. If the answer is in the affirmative, the research may be conducted, but there is no assurance that it will be.

This brings us to the second-tier of regulation. Once it has been determined that it is not socially irresponsible to permit persons to participate as subjects in a particular research project, we must then face the issue of how individuals are to be selected to participate (or are to be exempted from participating) in the research project. In theory at least, there are several possible mechanisms for selection and exemption.[4] A lottery might be used to choose particular subjects from the larger group of eligible subjects. Or authority could be delegated to the researchers to select subjects on the basis of medical criteria alone. Subjects might also be selected on the basis of non-medical social criteria. All of these methods of selection pose difficult ethical dilemmas.

The method of subject selection that we actually rely upon is the individual choice of those who wish to become research subjects. The subject's informed consent is the mechanism used to implement the principal of individual choice,[5] which is to say that those whose medical condition makes them fit for inclusion in a research project are provided with an opportunity to be included, but only if they so choose. This method of selection of individual subjects—self-selection, we might say*—runs the risk, unlike a lottery or conscription, that the research may not be performed at all, or may be performed with so few subjects as to be unable to yield statistically significant results. In effect, individual citizens are provided with a veto over the conduct of medical research. One stunning example of such a veto is recorded in the legal annals, involving psychosurgery research that had been approved by a review committee as being socially beneficial, but for which no subjects ever volunteered to participate.[6]

Informed Consent to Ordinary Medical Procedures

Before introducing the subject of incompetence, it will first be necessary to discuss informed consent in more detail in order to understand what incompetence is, and how it is related to the process of obtaining informed consent.

The requirement of informed consent originated not in the realm of medical research, but in conventional medical therapy, and it is in the latter realm that it has developed into the complex doctrine that it is today. The origins of the informed consent doctrine, itself, lie entirely outside of medical practice. Informed consent to treatment is an outgrowth of the earlier requirement that a doctor must obtain a patient's "consent" to treatment that is itself an outgrowth—or more properly, an illustration—of the ancient common-law protection accorded to bodily integrity by the law of trespass. If one were "touched" by another without consent, that touching constituted a trespass to the person, otherwise known as a battery. Even if no physical harm re-

*There is one important distinction between informed consent in the non-research setting and informed consent in the research setting. The notion that persons choose to be research subjects is slightly more attenuated, even in theory, than the notion that patients choose which therapeutic procedure, if any, to undergo. With the exception of normal subjects recruited through advertisements, most research subjects are patients of a physician who also wears the hat of researcher, or who is associated with another physician–researcher, and who are sought out for inclusion in the research—or "recruited," a particularly telling term.

sulted, the non-consensual nature of the touching made it a legal wrong for which redress might be obtained under a writ of trespass. Indeed, not only was absence of physical harm no barrier to legal redress, but a touching that benefited a person might be grounds for a lawsuit as long as it was non-consensual. There are several contemporary cases involving doctors and patients that illustrate this point.[7]

Thus, it is a fundamental precept of our legal system that every person has a right to bodily integrity, and correlatively, a right to decide when the right of bodily integrity is to give way to some other interest. This latter right, which I refer to as "decisional autonomy," has deep roots not only in the common-law tradition, but more recently has also found positive sanction in the constitutional right of privacy.[8]

The practice of medicine constitutes an interference with bodily integrity. However, because this interference is intended to be beneficial, individuals are often not merely willing to permit such an interference, but they affirmatively seek it out. In the medical context, the rights of bodily integrity and decisional autonomy are implemented through the requirement of informed consent. That is, before a medical procedure may be performed by a physician on a patient, the physician must obtain the patient's informed consent to treatment, by which the patient, in the exercise of his decisional autonomy, surrenders in a limited fashion his right of bodily integrity.

To obtain informed consent, a physician must do at least two things: inform the potential patient and obtain his consent.

Information Disclosure

Patients must be provided with all information "material" to making a decision whether to undergo or forego treatment.[9] This information must be provided by the physician or by someone to whom this task has been delegated, though the responsibility for seeing that it is properly done remains that of the physician. Patients need not be given information that they already know or which they can reasonably be assumed to know either by virtue of their own experience or by virtue of the fact that the information is common knowledge.[10] Among the kinds of things that the doctor must tell the patient are the material risks of treatment, the anticipated benefits, and alternative kinds of treatments.[11]

Consent

Whereas the problems associated with the requirement of information disclosure are largely of a practical nature, the issues associated with the "consent" requirement exist at both a conceptual and a practical level. About the only thing that is clear about consent is that the patient must give the doctor permission to perform the procedure, but even that is subject to some qualification. The case law is extremely unclear—and the two dozen informed consent statutes do not clarify the matter—as to whether anything more than the patient's mere permission is required. What could be required, in addition, is permission based upon understanding of the information that was disclosed—that is, understanding of the nature and consequences of the proposed medical procedures. What could also be required is that the doctor make reasonable efforts to determine whether the patient understands, and if the patient does not, to make further reasonable efforts to attempt to get the patient to understand.[12] If in the final anaylsis, the patient does not understand the information, it is uncertain as a legal matter whether or not the permission he gives provides the doctor with authority to perform the procedure, or even whether the patient's refusal is binding on the doctor.

Informed Consent to Research Procedures

Informed consent to research had its origins in American law primarily in the Nuremberg trials following World War II, in which several German physicians, in cooperation with their government, performed medical "experiments" on prisoners of war and concentration camp detainees. One aspect of the Nuremberg judgment, referred to as the Nuremberg Code, promulgates requirements for the ethical conduct of medical experimentation, one of which is informed consent.[13] Subsequently, the World Medical Association in its 1964 Declaration of Helsinki also subscribed to the requirement of informed consent to experimental procedures if the subject is competent, or from the "legal guardian" if the subject is not.[14]

It was not until 1966 that the US Public Health Service incorporated the substance of the Nuremberg Code and the

Declaration of Helsinki into guidelines for researchers that were then modified and published as the "Institutional Guide to DHEW Policy on Protection of Human Subjects" in 1971. This then became the basis for the DHEW regulations for the protection of human subjects, first issued in 1973,[15] and amended several times since then, most recently and thoroughly in 1981 by the DHHS.[16]

Whatever uncertainty there might be about the applicability of the common-law informed consent requirements to research procedures—and I suggest that there ought to be none—should be dispelled by the informed consent requirement mandated by DHHS regulations, which, however, are limited in their applicability to research supported by DHHS grant or contract. Under the regulations as revised in 1981, in order to obtain informed consent, an investigator must provide the subject with the following information:

> *1. A statement that the study involves research, an explanation of the purposes of the research and the expected duration of the subject's participation, a description of the procedures to be followed, and identification of any procedures which are experimental;*
>
> *2. A description of any reasonably foreseeable risks or discomforts to the subject;*
>
> *3. A description of any benefits to the subject or to others which may reasonably be expected from the research;*
>
> *4. A disclosure of appropriate alternative procedures or courses of treatment, if any, that might be advantageous to the subject . . .*[17]

In addition to these four requirements which closely follow the common-law requirements, investigators must also give subjects information about confidentiality, compensation, or medical care if the subject is injured, and a few other matters.[18]

Although there are some concrete differences between the common-law and regulatory requirements for informed consent, there are no conceptual differences. Both require 1. that relevant information be provided the patient–subject, 2. that consent be obtained, 3. that the patient–subject be so situated as to be able to render a voluntary decision, and 4. that the patient–subject be competent. Although there is no explicit requirement in the regulations that subjects be competent, this requirement is implicit in the statement that informed consent must be obtained from the individual or his legally authorized representative.[19]

The Presumption of Competency

The law presumes that all persons are able to exercise their right of decisional autonomy, that is, that they are legally "competent." Like the presumption of innocence that attaches to the criminally accused, the presumption of competency is not a matter of fact; not all persons are in fact innocent of crime, and not all patients in fact have the capacity to make medical decisions. Rather, a legal presumption is a device for instructing the authorities—courts in the one case, doctors in the other—as to how to proceed in the first instance.

In the criminal court, the import of the presumption of innocence is that the state must first come forward with some evidence of the accused's guilt; the accused is entitled to remain silent and make no move to defend himself. If the state fails to bring forth evidence of guilt, the accused is to be set free. This is in contrast to a converse presumption that would require the accused in the first instance to defend himself and to prove his innocence, or to remain imprisoned.

A similar result obtains in the medical context. An individual who comes to a doctor—indeed, even an individual who is brought to a doctor—is presumed competent to make decisions about his medical care. The patient has no burden to demonstrate that he has the capacity to make a decision; this is the meaning of the presumption of competency. Rather, to be deprived of his right of decisional autonomy, the physician must demonstrate that the patient lacks the capacity to make a decision. Failing this, the physician is bound by whatever decision the patient may make, however wrongheaded that the decision may seem to the physician.

For purposes of discussion, I have made one general assumption and two corollary assumptions that bear further scrutiny. My general assumption is that whatever analysis applies to informed consent and to incompetency in the ordinary physician–patient relationship also applies in the investigator–subject relationship.

The first corollary assumption of this is that the consequences of finding a potential research subject incompetent ought to be the same as the consequences of finding a patient incompetent. This is not necessarily so because although therapy may be assumed to be beneficial ex hypothesis, an experimental procedure may not stand the test of this assumption. Some research procedures are not even intended to benefit the subject.

The second corollary assumption is that research subjects are to be informed of the same categories of information that would be provided to a patient. Though this is generally the case, one necessary clarification or amplification is that the subject must also be informed that the procedure is experimental, and, where it is the case, that it is also not intended to be beneficial. Such information arguably is included within the requirement that the subject be informed of the "nature" of the procedure. Further, to the extent that the informed consent doctrine requires that the patient–subject understand the information that has been disclosed, the subject must understand that the procedure is experimental, that it may not be intended to benefit him, that it may not benefit him in fact, and that (where relevant) the choice of procedure is determined not by clinical judgment, but by random (or another non-discretionary method of) assignment.

Because an individual may be unable to engage in a discussion about information relevant to the research, or because in the course of a conversation the subject may not be able to understand the information, the problem of incompetency may actually arise in two different ways. Or, put somewhat differently, there may be two different kinds of incompetency—incompetency to be informed, and incompetency to consent.

Incompetency to be Informed: "Threshold Incompetency"

Statements abound in the judicial cases to the effect that only a competent individual may render consent for his own medical treatment. Thus, although the informed consent requirement obligates the doctor to make disclosure and obtain consent, these duties are suspended if the patient (or subject) is incompetent. Thus, incompetency is a condition, which if satisfied, calls into question the doctor's obligation to make disclosure and obtain consent. I will refer to this as "threshold incompetency," for if there is clear-cut evidence of incompetency at the threshold, the physician need not attempt to inform the patient and obtain his consent to treatment. Where there is less clear evidence of incompetency—such as where a subject is mentally retarded, but not profoundly so—this should serve to alert the physician–investigator that the presumption of competency that ordinarily prevails—that is, the presumption that the subject is entitled to make his own medical decisions—might not be operational in a particular case.

This, of course, does not mean that the physician is then free to render any treatment that may be necessary. Rather, in all cases except the most exigent,[20] the physician must obtain informed consent from the patient's surrogate, or, in the language of the federal regulations, the subject's "legally authorized representative." Furthermore, that a patient may be incompetent does not mean that he does not have preferences about his medical care nor that he is incapable of expressing them. Although there is no clear legal duty to do so, both the physician–investigator and the surrogate decision maker are ethically bound to consult with the patient, seek to elicit his preferences, and put them into effect insofar as is feasible in any given situation.

If the subject does not fail the test of threshold incompetency, the informed consent requirement obligates the doctor to make disclosure. If informed consent is viewed—as it should be—as something more than a simple stimulus–response model involving the input of information by the investigator into the subject and the spewing forth of a consent or refusal by the subject, the investigator will undoubtedly engage in a conversation with the subject. This conversation will involve, as most conversations do, a give-and-take of information, with the investigator telling the subject some things, the subject responding both verbally and behaviorally with indications of comprehension or confusion, with the subject occasionally asking questions of the investigator, and the investigator probably asking questions of the subject. If no further problems arise, the subject will then render a decision either to participate as a research subject or not. However, when dealing with subjects—the entire class of which is suffering from cognitive impairment—further problems are bound to arise.

Incompetency to Understand; Incompetency to Consent: "Process Incompetency"

In the course of having a discussion with a subject, the investigator may find that the subject understands little or nothing of what has been explained. Or, although the subject may understand in some sense, there may be something about the manner in which the subject uses the information that is highly idiosyncratic. Or, the prospective subject, even if he understands the information, may not be able to make a decision.

These kinds of events should encourage the investigator to probe more deeply to determine whether the subject understands what the investigator has disclosed. This probing may be accomplished by direct verbal questioning, or it may be done more subtly and indirectly. As long as the investigator maintains an interrogative posture toward the subject—that is, as long as he is on the look-out for whether the subject understands—the investigator is likely to obtain a feeling for the extent of the subject's comprehension. Thus, incompetency may become manifest in the process of, or after, disclosure and obtaining consent. I will refer to this as "process incompetency."

Both sorts of incompetency determinations are routinely made by physicians and investigators. The determination of threshold incompetency involves a general "sizing up" of the patient by the doctor. This may be done implicitly, and usually will be when the patient is clearly competent. However, in other situations gross features of the patient such as obvious alcohol or drug intoxication, obvious hallucinations or manifest delusions, serious mental retardation, or severe sensory disorders such as blindness or deafness will alert the doctor to the possibility, if not likelihood, that the patient is incompetent.

Incompetency

Cutting through the technical statement of the informed consent doctrine, what is clearly required is that the investigator engage in a conversation with the potential subject. Information is imparted to the subject; questions are asked; answers are given; more questions are asked and answered. All of this occurs, of course, so that the potential subject may determine in a reasoned manner whether or not to become a research subject.

Occasionally it may occur—indeed, it may occur frequently when certain subjects suffering from certain kinds of medical conditions are sought out—that the potential subject is unable to engage in a discussion about the research. Or, if the subject seems able to engage in a conversation, still it may be apparent to the investigator that the subject is not able to understand the information that is imparted. Or, although the patient may be able to understand the information, it may be clear that he is not able to use it, or does not in fact use it, in making a decision. Or finally it may be that the potential subject just does not make a decision.

In all of these situations, the presumption of competency to exercise decisional autonomy is called into question.

What Is Incompetency?

Up to this point, my discussion has assumed that the meaning of incompetency is self-evident. Though this is a convenient assumption for understanding the legal framework in which the concept is embedded, it will no longer suffice. We are, of course, talking about incompetency in a particular context: being a research subject. There are other kinds of incompetencies in law, each specific to a particular task or function: incompetency to stand trial, testamentary incompetency, or testimonial incompetency. The common thread of each is that the individual in question lacks the ability to perform the task at hand. One is incompetent to stand trial if one is unable to understand that he is to be tried for an offense, if he is unable to cooperate with his attorneys in the preparation of his defense, or if he is unable to appreciate the consequences of a conviction. One is incompetent to make a testamentary disposition if one does not know the natural objects of one's bounty or understand the nature and extent of one's estate. And one is incompetent to testify if one does not understand the obligation of the oath.

Since the task at hand is to make a decision whether to undergo or forego an experimental medical procedure, one ought to be considered incompetent if one is unable to understand the nature and consequences of and the alternatives to that procedure. In other words, one ought to be incompetent if one is unable to understand the information required to be disclosed by the investigator pursuant to the informed consent requirement.

I say that one ought to be considered incompetent under these circumstances because the law is lacking in certainty on this matter. There are extraordinarily few cases dealing with the incompetency to make medical decisions and none defining the term. In the few reported cases the person in question is undoubtedly incompetent because he is either unconscious or severely mentally retarded.[21]

Although incompetency denotes an incapacity of functioning in a particular manner, incompetency traditionally has been found to exist on the basis of certain statuses without regard to actual functionality. Certain statuses—children, the mentally retarded and mentally ill, and the intoxicated, for example—are associated with incompetency because as a general rule, persons occupying those statutes are non-functional or severely impaired in their functioning. However, though incompetency based on status is an indisputable administrative convenience, it is both functionally overinclusive and underinclusive in its sweep.

Although there may be a fairly good association between certain statuses and functional incompetency, it is best to attempt to ascertain incompetency directly on a case-by-case basis in order that persons who occupy a suspect status, but who are not actually incompetent, are not incorrectly determined to be, and vice-versa. Perhaps the best, and I might add only valid, use of status (except where the subject is a child) is to alert the investigator to the possibility of incompetency.

Functional tests of incompetency seek to answer the question "Is this patient able to participate in the medical decision making process in a meaningful way?" A negative answer to this question results in a finding that the patient is "incompetent," thus depriving the patient of decisional autonomy. In fact, this very question might be taken as a "test" of incompetency, except that it is so general as to be all but useless in particular cases. Rather, it is necessary to specify particular features of the medical decision making process which, if lacking, render the individual unable to participate in the process to the extent required by law. Functional tests of incompetency are unconcerned with the patient's status. Thus if a patient is, for example, "mentally ill," the presumption of competency is not automatically overcome. However, depending upon which functional test is utilized to determine incompetency, the effects that the mental illness has on the patient's cognitive abilities may be taken into account, and may, but need not necessarily, lead to the conclusion that the patient is incompetent.

Absence of Decision

One functional test of incompetency focuses on the absence or presence of a decision by the patient. A patient who chooses one treatment rather than another, or no treatment at all, is deemed competent. If the patient makes a choice, there is no further scrutiny of the manner in which he makes the choice, the reasons given for the decision, or the nature of the decision itself. By contrast, a patient who makes no choice when presented with the opportunity to do so is deemed incompetent. The mere failure to manifest a choice is determinative of incompetency. The person who is mute when asked to make a choice may well be incapable of receiving or communicating information, or such a person may be psychotic. If that is the case, this functional test of incompetency may overlap with a status test of incompetency. This test of incompetency allows the presumption of decisional autonomy to remain undisturbed unless there is extremely strong evidence of incompetency. Or to put it slightly differently, this test establishes an extremely high level of dysfunction as the test of incompetency.

Nature of Decision Making Process

Other tests of incompetency focus on the nature of the decision making process employed by the patient. After the patient is provided with the information mandated by the informed consent requirement, the doctor inquires into the manner in which the patient makes a decision concerning treatment. Certain ways of making decisions could be viewed as acceptable, and others as unacceptable. A patient who employs an unacceptable means of making a decision is thereby labelled "incompetent."

These approaches to the determination of incompetency are grounded in the view that if a patient is able to make a decision, but is unable to make it in the preferred manner, then the decision is something less of a decision and deserves less to be honored. The problem with this approach is that it is fundamentally inconsistent with the broad legal and ethical basis of the informed consent doctrine that permits patients to make decisions for their own idiosyncratic reasons if they so choose. Put another way, the doctor's duty of disclosure is intended to enable patients to make their decisions on the basis of the disclosed information, but not to require that they do so.

a. *Failure to Articulate Reasons in Support of the Decision.* A patient who is able to manifest a choice, and thus pass the "absence of decision" test, may still not be able to articulate reasons in support of that choice. Under this view, such a patient is deemed incompetent. This test would find more persons incompetent than the "absence of decision" test.

b. *Failure to Articulate Rational Reasons in Support of the Decision.* A person who could articulate a basis for his decision might still not be able to articulate rational reasons for that decision. That is, a patient might be deemed incompetent if the basis for the decision does not reflect both the information provided by the physician or other articulable reality-based information. This information need not necessarily be objectively factual; indeed the subjective value preferences of the particular patient such as his tolerance for pain and suffering, and his business, social, and personal obligations that might be compromised by treatment would all be legitimate reasons for a decision for or against treatment. By contrast, nonobjectively verifiable reasons—such as hallucinations or delusions—could be deemed nonrational grounds for decisions that would deprive the patient of his decisional autonomy. Needless to say, this test of incompetency is far more subjective than either of the foregoing tests.

c. *Failure to Employ a Utilitarian Calculus.* An even stiffer test of incompetency—that is, one which would deprive a far greater proportion of persons of their decisional autonomy—focuses on the patient's use of a utilitarian calculus to arrive at a decision. This test is suggested in the first instance by the informed consent requirement itself that, because it requires the doctor to disclose risks and benefits to the patient, could be construed as suggesting that the patient should weigh risks against benefits of treatment. A patient could easily articulate rational reasons in support of the decision that he makes, yet fail to weigh the benefits of a particular course of action against the risks. This test is even more subjective than the foregoing one because it not only requires the tester of incompetency to determine the factual veracity of a particular reason, but requires the judgment of the weight to be accorded to particular benefits and risks.

Nature of Decision

Incompetency could also be tested by reference to the outcome of the decision making process, rather than by reference to the nature of the process. For instance, the failure to make a decision that is in accordance with some externally verifiable standard might be deemed to render the patient incompetent. Examples of such standards are (a) what a reasonable person would decide under the same circumstances, or (b) what the physician has recommended. For example, any patient who chooses no treatment over treatment, or a risky treatment over a less risky one could be deemed incompetent if a hypothetical "reasonable person" would not make such a choice. Or a patient whose decision is different from the doctor's recommendation could be deemed incompetent. Such tests verge on undermining, if they do not actually undermine, the patient's decisional autonomy by honoring its exercise only where it is congruent with societal standards.

Lack of Understanding of "Informed Consent" Information

Another functional approach to incompetency involves determining whether or not the patient understands the information relevant to rendering an informed consent. There are two variants on this test:

a. *Actual Understanding.* The most straightforward way of applying this test is for the doctor (or other person)—who has made disclosure to the patient of the requisite information—to determine whether or not the patient understands it. A patient who does not understand the information is deemed incompetent, and deprived of his decisional autonomy. No inquiry need be made into how the patient uses the information or even whether he uses it; nor need there be any scrutiny of the reasons that the patient has for making a decision, nor of the nature of the decision itself. Rather, if the patient does not understand the information, he is deemed incompetent and deprived of his right of decisional autonomy.

A serious problem can occur in the administration of such a test from the fact that "understanding" is rarely if ever a simple "yes" or "no" matter. And further, since there is not merely one discrete bit of information that is disclosed but a range of information about risks, benefits, alternatives, and the nature of the procedure, as well as varying magnitudes and probabilities of risk and benefit, the measurement of understanding is a highly complex undertaking, to say nothing of establishing the level of adequacy of understanding.

This test best illustrates the conceptual overlap between incompetency and the "consent" element of informed consent. If consent means more than mere permission, as I earlier suggested that it does, and involves the giving of permission with an understanding of the nature and/or consequences of the touching that is to occur, then a requirement of competency is redundant. That is, when the courts state that a doctor may render treatment only on the basis of the informed consent of a competent person, they are either engaging in a redundancy, or they are requiring something else in addition to understanding of the disclosed information.

b. *Ability to Understand (Potential Understanding).* Instead of measuring directly the patient's understanding of the information given by the doctor, the patient's understanding of this information could be determined inferentially. The patient might be administered a formal intelligence test, for example. Or the patient's ability to understand informed consent information might be inferred from informal conversation with the patient. No matter what the basis of the inference, this variant encounters the same problems as the test based on actual understanding. Moreover, a further problem is introduced by the fact that the logical inference that is made may not be valid. This variant is similar to status tests of incompetency because it seeks to determine a patient's competency without directly measuring it, but instead by inferring it from something else.

Any test that seeks to determine "understanding" is particularly susceptible to the same problem that occurs with the "rational reasons" or "nature of decision-making" tests. In attempting to gauge understanding, the values of the tester play an insidious, though probably unavoidable, role. Not only does the tester's view of what constitutes understanding affect the determination of incompetency, but the initial selection of the information that the patient is to be tested on reflects the importance that the tester attaches to what information should be understood in order to be viewed as competent. Thus the personal identity and professional allegiance of the tester play a highly influential role in determining whether the patient is incompetent.

Who Decides Whether the Patient/ Subject Is Incompetent?

Regardless of what test of incompetency is to be applied, and regardless of how simple or difficult it may be to apply in particular cases, someone must apply it. That is, someone must determine whether a given research subject is competent or incompetent. Who is this to be?

Like the problem of choosing a test of incompetency, it is reasonable to assume that this is a settled matter. It is not. Only indirectly have the courts suggested that it is, in the non-experimental medical context, the responsibility of the physician to decide if the patient is incompetent. Although there are some practical and ethical difficulties with permitting the treating physician to have the first say on whether the patient is incompetent—especially where the patient is determined not to be incompetent and then undergoes a risky procedure (or foregoes a highly beneficial one because he has refused it)—without further judicial or legislative guidance, we may assume for present purposes that it is the treating physician's prerogative to determine incompetency in the non-experimental context.

Although the ethical problems are potentially more serious where the procedure is experimental—and especially where it is non-beneficial—there is at present no accepted alternative to having the investigator make the decision, short of a judicial determination that is complex, expensive, and time-consuming. Certainly where the investigator is also the subject's attending physician, it is best to seek consultation of another person before determining that the patient is incompetent. It may be even more important to seek consultation where the patient is determined not to be incompetent, but where competency is a close question.

Who Makes Decisions for the Patient/ Subject Who Is Incompetent?

Someone must make decisions about medical care (including participation in research) on behalf of an incompetent individual. Although there is no law expressly addressing decision making for incompetent prospective research subjects, a great deal of guidance can be obtained from a group of recent cases dealing with decision making about non-experimental procedures for incompetent patients.[22] On the basis of these cases—and in the absence of any express judicial or legislative guidance to the contrary—the most conservative course for investigators to follow is to permit incompetent patients to participate as research subjects only with the permission of a court-appointed guardian. A slightly less conservative position would be to relax the requirement of permission from a court-appointed guardian as the risks of the experimental procedures diminish and the potential benefits to the subject increase, but such a stance would provide less of a defense in the event of a lawsuit on behalf of a patient injured by a research procedure.

A question related to the one of who is to make decisions for incompetent patients is the discretion that such surrogate decision makers have in the kinds of decisions that they make. Surrogate decision makers are ordinarily guided by one of two legal standards: the "substituted judgment" standard or the "best interests" standard.[23] Under the former, the surrogate attempts to replicate the decision that the patient would make for himself if competent to do so. By contrast, the more objective "best interests" standard directs the surrogate to do for the patient what a reasonable patient would choose to do. Under either of these standards, however, it is not clear that a surrogate could consent to research procedures on behalf of an incompetent patient, unless the patient himself had expressed a wish to be the subject of such research prior to becoming incompetent.[24]

Conclusion

The legal status of conducting medical research procedures on patients with Alzheimer's disease or any other condition that calls into question their capacity to make decisions about the research is attended by a host of unanswered questions. Neither courts, nor legislatures, nor administrative agencies have provided explicit guidance on the critical issues of:

1. By what standards incompetency is to be determined
2. By whom is incompetency to be determined
3. Who is to make decisions for incompetent patients
4. The range of discretion that surrogate decision makers have.

Some guidance can be gleaned from cases involving questions of consent to non-experimental medical procedures, but because such procedures are presumed to be beneficial to their recipients while the benefit (and sometimes the risk) of experimental procedures is precisely the matter in question, the analogy between the two is an uncertain one.

Regardless, it is clear that investigators must obtain approval of research involving incompetent subjects—as they must with research involving competent subjects too—from an IRB, one of the duties of which is to take "appropriate additional safeguards . . . to protect the rights and welfare of . . . subjects . . . vulnerable to coercion or undue influence, such as persons with acute or severe physical or mental illness. . . ."[25] In carrying out this duty, an IRB may require the use of court-appointed surrogates to give permission for participation in research, and should take into account the potential risk to subjects from participation in comparison with the potential benefit.

Notes and References

[1]*. . . Some standard, by which to determine the propriety of treatment, must be adopted; otherwise experiment will take the place of skill, and the reckless experimentalist, the place of the educated, experienced practitioner . . . [When] the case is one as to which a system of treatment has been followed for a long time, there should be no departure from it, unless the surgeon who does it is prepared to take the risk of establishing, by his success, the propriety and safety of his experiment.*

The rule protects the community against reckless experiments, while it admits the adoption of new remedies and modes of treatment only when their benefits have been demonstrated, or when, from the necessity of the case, the surgeon or physician must be left to the exercise of his own skill and experience.

Carpenter v. Blake, 60 Barb. 488, 523 (N.Y. App. Div. 1871).

[2]*See* US DEPT OF HEALTH & HUMAN SERVICES, *Final Regulations Amending Basic HHS Policy for the Protection of Human Subjects*, 46 Fed. Reg. 8366, 8386–91 (January 26, 1981) [hereinafter cited as "Fed. Regs."].

[3]*See Ibid.* Section 46.111.

[4]G. CALABRESI & P. BOBBITT, TRAGIC CHOICES 41 *passim* (1978).

[5]*See* Fed. Regs. Section 46.116, *supra* note 2.

[6]*See* Kaimowitz v. Dept of Mental Health No. 73–19434–AW (Mich. Cir. Ct., Wayne Cty., July 10, 1973), reported in 1 MDLR 147 (1976).

[7]*See generally* Meisel, *The "Exceptions: to the Informed Consent Doctrine: Striking a Balance Between Competing Values in Medical Decisionmaking,* 1979 WIS. L. REV. 413, 415n.8 & 419.

[8]*See In re* Quinlan, 70 N.J. 10, 355 A.2d 647 (1976); *In re* Storar, 52 N.Y.2d 363, 420 N.E.2d 64 (1981); *In re* Spring, 405 N.E.2d 115 (Mass. 1980); Severns v. Wilmington Medical Center, Inc., 421 A.2d 1334 (Del. 1980); Superintendent of Belchertown State School v. Saikewicz, 370 N.E.2d 417 (Mass. 1977); Satz v. Perlmutter, 362 So.2d 160, 162 (Fla. Dis. Ct. App. 1978), *aff'd,* 379 So.2d 359 (Fla. 1980).

[9]In general the states are divided about how to measure the adequacy of information given to a patient. Some require that patients be given what a reasonable patient would find "material" to making a decision about treatment while others require the physician to provide patients with that kind and amount of information which reasonable members of the medical profession customarily provide. *See generally* Meisel & Kabnick, *Informed Consent to Medical Treatment: An Analysis of Recent Legislation,* 41 U. PITT. L. REV. 407, 421–426 (1980).

[10]*Ibid.* at 427.

[11]Fed. Regs. Section 46.116(a)(2)–(4), *supra* note 2.

[12]*See* Meisel, *The Expansion of Liability for Medical Accidents: From Negligence to Strict Liability by Way of Informed Consent,* 56 NEB. L. REV. 51, 113–123 (1977).

[13]*See* J. KATZ, EXPERIMENTATION WITH HUMAN BEINGS 305–306 (1972).

[14]*See* 271 N. ENG. J. MED. 473 (1964).

[15]US DEPT OF HEALTH, EDUCATION & WELFARE, *Protection of Human Subjects,* 39 Fed. Reg. 18,914 (May 30, 1974).

[16]Fed. Regs., *supra* note 2.

[17]*Ibid.* Section 46.116(a)(1)–(4).

[18]*Ibid.* Section 46.116(a)(5)–(8).

[19]*Ibid.* Section 46.116(a).

[20]Informed consent to non-experimental medical procedures need not be obtained in an emergency. *See* Meisel, *The "Exceptions" to the Informed Consent Doctrine, supra* note 7, at 434–438. Whether or not experimental medical procedures may be performed in an emergency without obtaining informed consent is unclear. Section 46.116(f) of the Fed. Regs., *supra* note 2, purports to deal with this matter, but does not give any useful guidance. *See* Abramson, Meisel, & Safar, *Informed Consent to Resuscitation Research,* 246 JAMA 2828 (1981).

[21]One of the few exceptions is State Dept. of Human Services v. Northern, 563 S.W.2d 197 (Tenn. Ct. App. 1978). In that case, the patient was able to communicate but was found incompetent nonetheless because of the inconsistency and illogic inherent in the directives that she was giving to

her caretakers. The court failed to provide a universal or even a general standard by which incompetency is to be determined, confining itself narrowly to the facts of the case, and stating:

> *[T]his Court has found the patient to be lucid and apparently of sound mind generally. However, on the subjects of death and amputation of her feet [which were the probable consequence and proposed procedure respectively], her comprehension is blocked, blinded or dimmed to the extent that she is incapable of recognizing facts which would be obvious to a person of normal perception.*

Ibid. at 209–210. What the court says amounts to little more than the patient is incompetent because she is not normal, hardly a helpful guide.

[22]*See In re* Storar, 52 N.Y.2d 363, 420 N.E.2d 64 (1981); *In re* Spring, 405 N.E.2d 115 (Mass. 1980); Superintendent of Belchertown State School v. Saikewicz, 370 N.E.2d 417 (Mass. 1977); Custody of a Minor, 385 Mass. 697, 434 N.E.2d 601 (1982); *In re* Dinnerstein, 380 N.E.2d 134 (Mass. Ct. App. 1978); *In re* Roe, 421 N.E.2d 40 (Mass. 1981).

[23]*See* Superintendent of Belchertown State School v. Saikewicz, 370 N.E.2d 417 (Mass. 1977).

[24]*Cf. In re* Storar, 52 N.Y.2d 363, 420 N.E.2d 64 (1981).

[25]Fed. Regs. Section 46.111(b), *supra* note 2.

Assessment of Competence to Give Informed Consent

Allen R. Dyer

Introduction

The idealized conception of informed consent requires a so-called "autonomous," rational, and often highly intelligent person who is capable of understanding the research in which he or she is asked to participate and who possesses the freedom to make an uncoerced choice about whether to participate. Given the complexity of both contemporary research and of human motivations, are even rational and sane people really capable of giving a truly informed consent? Psychiatric patients, children, prisoners, the senile or demented, the comatose, and participants in research involving deception all lack the inner freedom required for a truly informed consent. Must they therefore be excluded from research that ultimately might be of benefit to them or to society?

Clearly not, but we must look closely at our concept of informed consent in order to understand how best to protect both the rights and interests of those whose capabilities are limited. If we think about informed consent as an action of the autonomous person, we are left at an impasse about how to treat the nonautonomous person. The customary approach of defining the nonautonomous person as also noncompetent is probably more drastic than necessary. After sketching the contours of the problem, I will propose a way of thinking that relies more on the relationships between people than it does on the notion of an isolated and autonomous person.

Informed Consent Criteria

Strictly speaking, informed consent has three components; it must be informed, voluntary, and competent. In this country we usually speak merely of "informed consent," belying our undue emphasis on the information component instead of the human process of consenting. To state my argument as succinctly as possible: I believe that our Western intellectual traditions have led us dangerously close to equating personhood with rationality, and that overly legalistic notions of informed consent actually work to the detriment of persons with diminished capabilities—and indeed to the detriment of all of us—by emphasizing criteria that divide us from one another and may exclude some from participation in human communities. Furthermore, overreliance on the ability to understand "information" may actually do more to protect a cherished notion of rationality than to safeguard potential research subjects.

Antonomy vs Paternalism

A cultural conflict, embedded in the very notion of informed consent, contrasts two visions of human beings and their relationships to one another. This conflict seems to be at the very center of Anglo-American law and ethical theory. It contrasts a vision of human beings as autonomous persons, and yet maintains a deference to paternalism by declaring the nonautonomous person as incompetent, hence in need of paternalistic intervention. The conflict created by uncertainties about the extent to which individual and societal well-being is better served by encouraging self-determination or supporting paternalism is central to the problem of informed consent. This fundamental conflict, reflecting a thoroughgoing ambivalence about human beings' capabilities for taking care of themselves and wishes for dependency, has shaped our thinking about informed consent more decisively than may be commonly appreciated.

Community and Consent

"Consent" in the dictionary definition derives from the Latin *com* + *sentire:* to feel; hence, to feel together. It means "agree," "assent," or "give permission and indicates involvement of the will or feelings and compliance with what is requested or desired." Implicit in this definition of consent is a community of feeling—a shared trust. With recognition of this mutuality, Otto Guttentag offers a useful definition:

> *"Informed consent may be defined as the experimenter's willing obligation to inform the experimental subject, to the best of the experimenter's knowledge, about the personal risk that the experimental subject faces in the proposed experiment, the significance of the experiment for the advancement of knowledge and human welfare, and last but not least, the stakes involved for the experimenter himself.*[1]

Following this very personal view of ethics, Guttentag offers the notion of "partnership" as the basic ethical principle in experiments involving human subjects. "It is the concept of partnership between the two, resulting from the fact of their being fellow human beings, that reflects our basic belief and cannot be subordinated to any other." By partnership he does not imply a legal contract or business association, but rather "sincerity without reserve," a relationship of mutual trust and confidence, or openness between investigator and subject, and a blind reliance that discards any guardedness.

Autonomy and Consent

An alternative contender for the title of "fundamental ethical principle" is the principle of autonomy. Robert Veatch argues that "the most plausible foundation for informed consent is the principle of autonomy."[2] This principle holds that individuals are the possessors of individual rights, including the right to self-determination. In this view, we require informed consent because the autonomous person has a right to decide for himself or herself.

The principle of autonomy has the merit of bolstering against possible violations of trust. It has the liability of degenerating to an abstraction, and of being applied to groups without due regard for the idiosyncrasies of individuals. Fundamentally, the principle of autonomy makes sense not as an abstraction, but as an extension of the fiduciary principle. Autonomy becomes meaningful not in isolation, but only when it is respected by other persons.

Though there is widespread acceptance of the idea that the principle of autonomy is indeed the basis of our requirement for informed consent, I am concerned that our preoccupation with "autonomy" as an absolute has become an obsession reaching pathological proportions, allowing us to substitute the idea of a rational, independent mind of a Cartesian sort for the painful realities of being sick, dependent, and in need of help.

Acknowledging the very legitimate concern about potential abuses of power in paternalistic relations, I suggest that we not be seduced by a fruitless adversity between paternalism versus autonomy, but rather that we specifically concern ourselves with the dynamics of dependency relationships. When can people function more or less independently, and when and how must they best be cared for?

The Assessment of Competence

If we rely on the principle of autonomy as the basis for our requirement of informed consent, we are faced with problems about what to do with those persons who cannot be said to be competent. Taking a strict legal interpretation of informed consent, the right to self-determination cannot be exercised without the ability to understand the information. This creates a conflict with the exercise of another right—the right to participate in activities that might be of benefit to the self or society. This conflict is usually resolved or avoided by declaring the nonautonomous person incompetent, or by understanding competence broadly.

Tom Beauchamp and James Childress make the following useful observation, one that demonstrates the complexity of considerations necessary to maintain the primacy of the principle of autonomy:

> *The concept of competence is a multidimensional one. Competence and incompetence are often assessed by diverse and even inconsistent theories of comprehension, rationality, freedom, physiological states, . . . and judgments of incompetence often apply to a limited range of decision making, not to all decisions made by a person. Some persons who are legally incompetent may be competent to conduct most of their personal affairs, and vice versa. The same person's ability to make decisions may vary over time, and the person may at a single*

> *time be competent to make certain practical decisions but incompetent to make others. For example, a person judged incompetent to drive an automobile may not be incompetent to decide to participate in medical research, or may be able to handle simple affairs easily, while faltering before complex ones. Accordingly, the notions of limited competence and intermittent competence are useful, because they require a statement of the precise decisions a person can make, while avoiding the false dichotomy of "either competent or incompetent." Use of these notions preserve maximum autonomy, justifying intervention only in those instances where a person clearly is of questionable competence.*[3]

Competency is a legal concept and all individuals are presumed by law to be competent until determined otherwise by a judicial hearing. The practical realities of clinical care often require an assessment of competence to refuse or consent to a particular procedure. Paul Appelbaum and Loren Roth stress the dynamic qualities of competence for which static legal theories do not make adequate provision. They draw a distinction between "psychological capacity" and "legal competence."[4] They suggest five criteria that should be used in assessing competence:

1. Psychodynamic elements of the patient's personality;
2. The accuracy of the historical information conveyed by the patient;
3. The accuracy and completeness of the information disclosed to the patient;
4. The stability of the patient's mental status over time;
5. The effect of the setting in which consent is obtained. These criteria help broaden the concept of competence to be responsive to the needs of a particular person at a particular time.

The Perception of Knowledge

Here, a brief digression into epistemology is in order. Our concept of informed consent and our tendency to rely on the principle of autonomy to justify it are based on an often unexamined conception of knowledge. Even though there is a tendency to equate them, knowledge and information are not the same thing. That tendency arises from the notion that science requires knowledge to be precisely specifiable and admits no place for ambiguity or unspecifiability. Science, and hence, knowledge in this view are "objective."

Various philosophers of science, such as Thomas Kuhn,[5] Karl Popper,[6] and Michael Polanyi,[7] have challenged this view of scientific objectivity by looking at the way new theories come to be accepted by the scientific community. Scientists, who are human, must become convinced of the validity of new theories, and this process is not always "consensual, as is often held, relying on predetermined criteria of what is acceptable evidence. Rather it relies on much discussion and debate, and on standards of which even scientists may not be consciously aware. Polanyi, for example, draws on the work of Gestalt psychologists to illustrate what is at stake for a scientist in pursuit of discovery. "We know more than we can tell, and in order to specify what we know, we must rely on our awareness of things that we may not be able to specify."

Polanyi uses the familiar optical illusions of the Gestaltists to demonstrate the basis for perceiving and knowing quite generally; two opposing silhouettes become a goblet by a slight change in perspective (Figure 1), or two lines of equal length appear to be different lengths when arrows pointing in opposite directions are drawn on these lines. In each of these instances there is a mutually exclusive but related awareness of the figure and its ground. The perception may be changed by altering the focus to the figure as ground and vice versa.

The point may be more easily grasped metaphorically and symbolically by contrasting the Japanese language and modes of thought with the Anglo-American. Figure 2 shows three

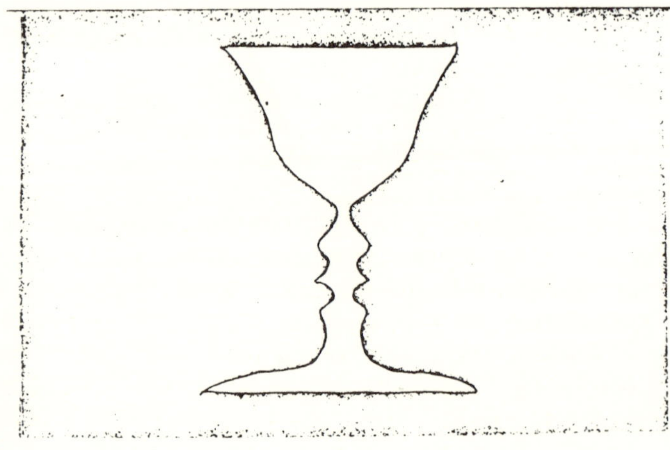

Fig. 1

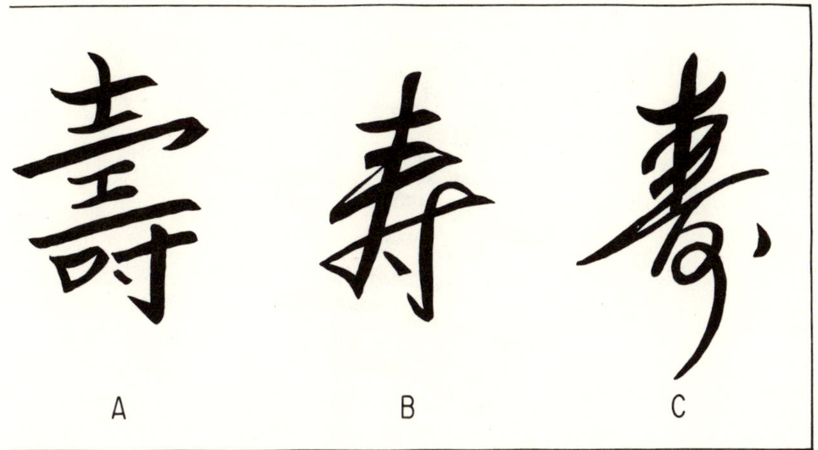

Fig. 2

versions of the Japanese character meaning "lucky" or "congratulations."[8] On the left is a formal version—used for special greetings such as wedding invitations—while on the right are simpler and more commonly used versions of the same character. Japanese children learn to recognize the meaning conveyed by this symbol by a kind of pattern recognition; they perceive the pattern instantly, rather than by a line-by-line analysis of the way it is constructed. Thus the three forms are perceived to convey the same meaning, though they are constructed differently.

A binary computer, however, recognizes the pattern by a very different process. (Figure 3) A fine grid is superimposed over the character, and each block is checked to see if it is black or white. The character is then defined by a larger number of "on or off," "0 to 1," or "yes or no" determinations.

The analogy to the question of determining competence is clear; we are presented with two alternative models of understanding. One is precisely specifiable, but inflexible; the other is readily recognizable, but variant and ambiguous. The determinations of competence or the validity of consent requires more than laying a template of guidelines on a particular person or situation. It requires an empathic understanding of another person that cannot be completely specified. If philosophers of science such as Popper, Kuhn, and Polanyi are correct, I am not merely contrasting Eastern and Western modes of perception, because Western science—the scientist as a person in pursuit of discovery—actually proceeds much more in the fashion of the

Fig. 3

Japanese child than of the binary computer. I suggest that our legal and ethical modes should strive for a similar fidelity to human experience, rather than an abstract notion of precision, objectivity, and standardization.

Beyond the Impasse

A so-called "autonomous" person may be able to make a free choice, given enough information about the proposed research. The nonautonomous person may lack the freedom to make such choices independently. Although our IRB procedures tend to focus on the information in the consent forms, the act of obtaining an informed consent actually involves us in two contradictory activities simultaneously. On the one hand, we strive for a conceptual clarity in the full disclosure. On the other hand, we abandon such abstract clarity on behalf of the person who must understand the consent form. (The form must be understandable.) One movement stresses the information in the consent form. The other movement stresses the process of consenting.

Solutions to these alternatives that focus on the "information" aspect of the informed consent are inadequate. An examination of the subjects on the contents of the form would disclose more about test-taking skills than the subjects' intentions. A categorical exclusion of patients with diminished competence would ensure against violations in the informed consent, but in itself risks violations of certain rights such as the right to participate in the search for knowledge about one's own illness. The traditional "proxy consent" substitutes another person for the research subject—person who presumably is more capable of understanding the information in the consent form, but who may or may not be in a good position to speak for the interests of the patient–subject. As long as our criteria of adequately informed consent rest on the information transmitted, only a case-by-case review will strictly satisfy our demands; however, this is cumbersome.

Since trust cannot be assumed, it becomes necessary to separate the roles of the investigator and the person who actually obtains the informed consent.[9–10] If we keep in mind the simultaneous movements in obtaining an informed consent, the basis for separation becomes evident. The patient–subject does not negotiate the informed consent directly with the investigator, who has a dual interest not only in the patient as a person, but also in the outcome of the investigation. Rather, the patient–subject negotiates with a third person, whose task is to represent the patient's interest and speak for the patient if the patient is in need of such help. I prefer the term "trustee" for this position to emphasize the fiduciary nature of the relationship. It is the trustee who would negotiate with the investigator, and their relationship could be as information-oriented as necessary to clarify what should be included in the consent form.

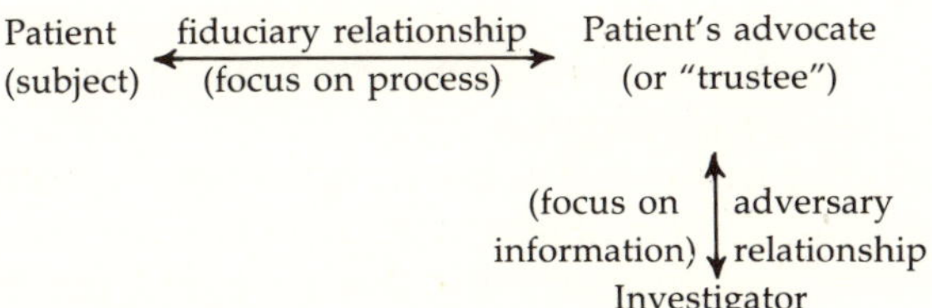

The patient's advocate might well be the attending physician—the physician–friend—if this were not also the physician–scientist. Other candidates for this role would vary depending on the setting. A close relative or legal guardian might be able to speak in some instances for his or her relative or ward. A nurse in other settings might be the person who would best know the interests of the patient. In a research institute or state mental hospital, a specially trained lay person (paid or volunteer) might assume the role of the patient's advocate.

The key to this system hinges on the integrity of the person who serves as the patient's advocate or "trustee." That person might be capable of understanding and speaking for the patient's interests, and be willing to spend time with the patient to understand what those interests might be. Empathy would be the key criterion. In practice, patients' advocates have often been strict civil libertarians, who hated [paternalistic] physicians and have seen freedom as the only goal worthy of consideration. Such people are simply not qualified to assume the responsibilities of speaking for another person! The responsibilities that go with such trusteeship are subject to the same liabilities of paternalism that plague anyone in positions of power and responsibility.

Another merit of this two-step system is that it offers a second level for review that is practically more feasible than reviewing each signature and each transaction. The IRB would continue to review the content of the consent form. The process of obtaining the consent for nonautonomous persons could be the subject of a different kind of review that could be conducted by either the IRB or by a site visit. The second review would focus on a process to determine that the transaction of the informed consent was being handled properly. One feature of this review might be, for example, to determine who might best serve the role of patient's advocate in the particular setting, and which groups of nonautonomous subjects might be in need of such trusteeship for particular kinds of research. Another important task would be to assure that those who assumed the important role of advocate were capable of listening to and speaking for the patient, not merely articulating their own political commitments or ideologies.

The notion of informed consent has evolved into a central position in our medical and scientific procedures involving human beings. It serves to focus our thinking and practice on the ethical dimensions of our activities. It is the focus on informed consent that forces us to rethink our value of rationality, our tradition of intellectual clarity, the ambiguities attendant to communication through the symbolism of language, and how we deal with ambivalent affects. In so doing we might decide to disenfranchise those dissimilar from ourselves, or we might come to recognize something of ourselves in the nonautonomous persons with whom we are concerned. This may be a humbling experience, especially if our concept of ourselves has tended to stress the cognitive side of our being, but it may be a humanizing experience as well. The view of informed consent stressed here emphasizes the relatedness of people more than their autonomy. This approach brings us into closer proximity with human suffering than may be comfortable, but it seems at least a minimum ethical requirement for participation in those procedures for which we have come to expect an informed consent.

Notes and References

[1]O. Guttentag, "Ethical Problems in Human Experimentation," *Ethical Issues in Medicine,* E. Fuller Torrey, ed. (Boston: Little-Brown, 1968).

[2]R. Veatch, "Three Theories of Informed Consent: Philosophical Foundations and Policy Implications," *The Belmont Report: Ethical Principles and Guidelines for the Protection of Human Subjects of Research* (Washington, DC: DHEW Publication No. (OS) 78–0014, 1978, pp. 26.1–26.66.

[3]T. L. Beauchamp and J. F. Childress, *Principles of Biomedical Ethics* (Oxford University Press, 1980), pp. 67–68.

[4]P. S. Appelbaum and L. H. Roth, "Clinical Issues in the Assessment of Competency," *American Journal of Psychiatry* 138, 1981, 1462–1467.

[5]T. Kuhn, *The Structure of Scientific Revolutions* (Chicago: University of Chicago Press, 1962).

[6]K. Popper, *The Logic of Scientific Discovery* (New York: Basic Books, 1959).

[7]M. Polanyi, *Personal Knowledge: Towards a Post-Critical Epistemology* (London: Routledge and Kegan Paul, 1952).

[8]M. Kikuchi, "Creativity and Ways of Thinking: The Japanese Style," *Physics Today,* 1981, 42–51.

[9]Guttentag, O. "The Problem of Experimentation on Human Beings: THe Physician's Point of View," *Science* 117, 1953, 207–210.

[10]A. R. Dyer, "Can Informed Consent be Obtained from a Psychiatric Patient?" *Controversy in Psychiatry,* J. P. Brady and H. K. H. Brodie, eds. (Philadelphia: W. B. Saunders, 1978), pp. 983–996.

Part 6

Proxy and Derived Consent

Autonomy and Proxy Consent

Bruce L. Miller

Introduction

A necessary condition for ethical research on humans is the consent of the subjects. Without consent, the subjects are used by the researcher for the good of others, and though risks to the subjects may be slight and believed to be offset by the possible benefits to others, the subjects' right to autonomy or self-determination is violated. If the subjects are competent and their informed consent is obtained, the benefits of research can be realized without compromising the subjects' right to autonomy. When the subjects in research are not competent, or are partially and intermittently competent, as is likely in research on Alzheimer's disease, protecting their right to autonomy becomes problematic.

A proposed solution to this problem is to require "proxy consent" when the subject is not competent or as the DHHS regulations state: "Informed consent will be sought from each prospective subject or the subject's legally authorized representative."[1] DHHS assumes that the rights of noncompetent or partially competent subjects can be protected, but leaves the determination of how this is done to IRBs. In its reports on research on children and on the institutionalized mentally infirm,[2] the National Commission was more specific on the process of proxy consent. The Commission abandoned use of the term "proxy consent" on the grounds that proxy consent was not consent at all. Instead they speak of the "permission" of a guardian. The Commission placed restrictions on what a guardian could consent to by limiting the amount of risk a subject could be exposed to, by requiring benefits to the subject or others, by the use of consent auditors, by requiring review by a national board in some cases, and by requiring "assent" of the subject when possible. These requirements may be offered as protections of the right to autonomy of subjects; they also seem designed to require a more favorable risk–benefit ratio for noncompetent subjects than for competent subjects. The result of this is that the right to autonomy may be sacrificed for noncompetent subjects in a way that it is not sacrificed for competent subjects, since research on competent patients requires informed consent regardless of the risk–benefit ratio.

There are different reactions to this. First, following Paul Ramsey,[3] one could say that procedures that hold out no prospect of benefit to the noncompetent subject are not morally permissible, and that the Commission's recommendations should be rejected in favor of a more stringent approach. Second, one could follow Richard McCormick[4] and argue that everyone—competent and noncompetent alike—has a natural obligation to benefit society, and thus research that falls within the scope of that obligation is permissible. Thirdly, one could take the position that the right to autonomy is an important consideration for the ethical permissibility of research, but it is not the only value; and that in some situations, autonomy may be sacrificed, provided that it is slight and cautious. Each of these positions has its problems; Ramsey's position would prohibit needed research in diseases like Alzheimer's; McCormick's position provides no clear guidelines for when research participation falls within the natural obligation to benefit society; the last position, which seems the position of the National Commission, has the problems of a slippery slope because there are no clear guidelines on what the required risk–benefit ratio must be.

A better approach is to determine whether, and if so, to what extent, proxy consent is consistent with the right to autonomy. This requires a detailed analysis of the two concepts.

Autonomy

The first level of distinctions that must be made are between the three ways in which the term autonomy can be predicated:

1. Autonomy as a feature of particular actions of persons, including decisions
2. Autonomy as a feature of a person's capacities
3. Autonomy as a feature of interpersonal and institutional environments.

These three uses of "autonomy" are not unrelated, yet they are distinct. A person's action may not be autonomous because the individual acted impulsively, and did not consider the alternatives and their consequences in a manner appropriate to the situation; yet, the person may be autonomous in the sense of having the relevant capacity or in the sense of not being coerced. The person could have deliberated on the alternatives and their consequences but did not.

The second and third uses of autonomy are distinguished on the rough difference between internal and external limitations on the options available to a person. People may steal because they have neurotic compulsions to do so (kleptomania), because they see stealing as the only way not to starve, or because they are coerced by others to do so. If in each case there is a long term pattern of stealing, then we might judge that each person lacked autonomy in regard to stealing. The kleptomaniac is internally compelled to steal, the person coerced to steal is externally compelled, and the person who steals not to starve is probably compelled by a combination of external and internal factors.

Both internal and external limitations on options can lead to doubt about the adequacy of consent. A person whose cognitive ability is somewhat impaired by Alzheimer's disease may not be able to give sufficient attention to alternative treatments or alternatives to participation in research, though the person can know what he or she is doing. A person who is institutionalized and believes that he or she must cooperate with the staff, or lose certain privleges, may consent to whatever procedures are suggested by the staff without considering the alternatives. Not only may some limitations on options be a combination of internal lack of capacity and external restraints, but some cases of lack of capacity are the result, in small or large measure, of external factors. For example, a common cause of senile dementia, not of the Alzheimer's type, is the moderate depression that results from the isolation and lack of stimulation of the elderly.[5]

Not every incapacity and external restraint constitutes a lack of autonomy. From a purely descriptive point of view, persons could be compared in terms of their capacities and the external restraints on their options, and then the person with more or greater capacities and fewer external restraints would be more autonomous. Autonomy is just a matter of degree on this view. This confuses autonomy with the related, yet different, notions of power and independence. A person who has average physical and intellectual abilities and who is subject to the normal range of legal, social, and interpersonal restraints, is an autonomous individual.

When we say that someone lacks autonomy, e.g., a patient with Alzheimer's or the institutionalized elderly, we are not simply describing a relative lack of ability and opportunity. We are also making a moral claim that some special regard is due that person; that we ought to treat the person with more care and concern than we give to others; that we ought to do something about the lack of autonomy; or that the lack of autonomy has implications for what we can reasonably expect of the person. To say that a person lacks autonomy carries the implication that the individual is below some minimum level of capacity and range of opportunity, and not simply that the person has a lack of capacity and opportunity relative to some more powerful person.

This discussion of the autonomy of persons as a feature of a person's capacities and of a person's environmental restraints is very sketchy; a full account would require much more explanation. The purpose was simply to distinguish them from the notion of autonomy of actions.

Four Senses of Autonomy of Action

The four senses of Autonomy of Action are:

1. Autonomy as free action
2. Autonomy as authenticity
3. Autonomy as effective deliberation
4. Autonomy as moral reflection.[6]

Autonomy as free action means an action that is voluntary and intentional. An action is voluntary if it is not the result of coercion, duress, or undue influence. An action is intentional if it is the conscious object of the person to do that action. In the current context it is important to observe that it is an action to submit oneself, or refuse to submit oneself, to medical treatment or research. If it is the conscious object of a person to be a subject in a particular research protocol, and he or she submits to the research, the action of the person is intentional. If a person does not wish to be a subject and refuses to participate, that is an intentional action. If a person agrees to what is offered as a medical treatment, but is in fact entered into a research protocol that offers no medical benefit to that person, the person voluntarily submitted to the procedure, but it was not a free action because he or she did not intend to participate in research. The doctrine of consent—as it was before the law gave us the doctrine of informed consent—required that permission be obtained from a person and that the person be told what sort of procedure would be done; this maintains the right to autonomy as free action. Permission makes the procedure voluntary, and knowledge of the procedure makes it intentional.

Autonomy as authenticity means that an action is consistent with the attitudes, values, dispositions, and life plans of the person. The rough idea is that of acting in character. Our inchoate notion of authenticity is revealed in comments such as "That's just what I'd expect George to do!" or "That's certainly in character for her!" On the inauthentic side we say things such as "That's not the Jane Smith I know!" or "He's not himself today!" It will not always be possible to label an action authentic or inauthentic. On the one hand, a given disposition may not be sufficiently specific to judge that it would motivate a particular action—for example, a generous person need not contribute to every cause to merit that attribute. On the other hand, most people have dispositions that conflict in some situations; an interest in and commitment to scientific research will conflict with fear of invasive procedures when such an individual considers being a subject in medical research. There are many questions about this sense of autonomy that cannot be explored here, but the present account shows that there is something in our experience that can be characterized adequately as authenticity, and that it is part of the notion of autonomy.

Autonomy as effective deliberation is an action taken when a person believes that he or she is in a situation calling for a decision, is aware of the alternatives, aware of the consequences, and chooses an action based on that evaluation. Effective deliberation is of course a matter of degree; one can be more or less aware and take more or less care in making decisions. Effective deliberation is distinct from authenticity and free action. A person's action can be voluntary and intentional and not result from effective deliberation, as when one acts impulsively. Further, a person who has a rigid pattern of life acts authentically when he or she does the things we have come to expect, but without effective deliberation. The doctrine of informed consent, which requires that the patient or subject be informed of the risks and benefits of a proposed treatment and its alternatives, protects the right to autonomy when autonomy of action is conceived as effective deliberation.

Autonomy as moral reflection means acceptance of the moral values one acts on. The values can be those one was dealt in the socialization process, or they can be values that differ in small or large measure from the former. The important point is that one has reflected on them and now accepts them as one's own. This way of putting it is more plausible than saying that one makes one's own values, for that suggests that a person can separate himself or herself from social influences. This sense of autonomy is deepest and most demanding when it is conceived as reflections on one's complete set of values, attitudes, and life plans. It requires self analysis that not many are interested in, awareness of alternative sets of values, commitment to a method for assessing them, and the ability to put them in place. Occasional, or piecemeal, moral reflection is less demanding and more common. It can be brought about by a particular moral problem, and only requires reflection on the values and plans relevant to the problem. Autonomy as moral reflection is distinguished from effective deliberation because one can do the latter without questioning the values on which one bases the choice in a deliberation. Reflection on one's values may be occasioned by deliberation on a particular problem; so in some cases it may be difficult to sort out reflection on one's values and plans from deliberation using one's values and plans. Moral reflection can be related to authenticity by regarding the former as determining what sort of person one will be, and in comparison to which one's actions can be judged as authentic or inauthentic. Autonomy as moral reflection is the sense of autonomy most discussed by philosophers, but it is the sense of autonomy least relevant to questions about the ethics of research.

Much more needs to be done for a full account of these four senses of autonomy, including a more detailed explanation of each and further specification of their relationships. I hope that enough has been said to make the ideas sufficiently clear and persuasive so that their application to the ethics of research, and particularly to the notion of proxy consent, will be useful.

With these four senses of autonomous action clarified, it can now be seen that an inference from lack of autonomy as lack of capacity or as external constraint must be made with attention to what aspect of autonomous action is made unlikely, or impossible, by what specific lack of capacity or external constraint. For example, if a person's cognitive incapacity involves loss of short-term memory, that may make autonomy as effective deliberation impossible, but need not make autonomy as free action or authenticity impossible. Similarly, a person who is emotionally labile or depressed may not act autonomously in the sense of authenticity, but the person's actions are nonetheless autonomous as free actions. A person whose cognitive impairment includes loss of orientation to a great degree may not be able to act autonomously in the sense of free action; if a person does not know where he is, why he is there, and who he is with, it is not plausible to say that what he is doing is intentional. Concerning external restraints and autonomy of action, the most discussed issue is whether an institutionalized person can act autonomously. If an institution is structured so that persons in it cannot get the decent amenities of life without engaging in certain actions, doing those things can be autonomous in the sense of effective deliberation, and though they can be intentional, they are not voluntary; therefore, they are not autonomous in the sense of free action. If someone limits the amount of information another person can have, the action of the latter can be autonomous in the sense of free action and in the sense of authenticity, but not autonomous in the sense of effective deliberation. So whether an incapacity or external restraint makes autonomous action impossible depends on what aspect of autonomy of action one wishes to realize or protect, and whether the particular incapacity or external restraint interferes with it.

Models of Proxy Consent

There is nothing surprising about the observation that when one person acts as proxy or agent for another, there can be great variation in the specific circumstances that effect the acceptability or quality of the proxy action. Yet there are no widely used taxonomies of proxy consent in the medical ethics literature. I have constructed six different models, or ideal types, of proxy consent: specific authorization; general authorization with instructions; general authorization without instructions;

instructions without authorization; substitute judgment; and deputy judgment. The names are not being used in a way that purports to comply with some standard meaning; they are partly descriptive and partly arbitrary.

In describing these six models of proxy consent, the lawyer's terminology of principal and agent will be misused throughout. A principal will be the person who authorizes another to act, or on whose behalf another acts; an agent is the person authorized by the principal to act or who acts on behalf of the principal without the principal's authorization, but with the authority of some law or social custom. A central assumption is that on the occasions when the agent acts, the principal is not fully competent to take action on the matter of concern. Another assumption is that the action to be taken by the agent is to grant or deny permission that some other person, a physician or researcher, perform some medical or other research procedure on the principal. These assumptions limit the focus to proxy consent to therapy or research.

Specific Authorization

This is the situation where a principal appoints an agent to give or deny permission to some specific procedure at some future time when the principal anticipates that he or she will not be competent. Imagine a patient who has a terminal illness and knows it, and who instructs his or her spouse, friend, or other person not to consent to a respirator if and when the patient becomes incompetent. In the area of research, a principal may know that he or she has a disease that is likely to render him or her incompetent at some point, and specifically instruct another to give permission to a researcher to include the patient in a particular research protocol on that disease. Obviously, specific authorization can be done in conjunction with prior consent or refusal of the patient or subject. The patient in the first case mentioned above could have explained to the health care providers that he or she did not want to be placed on a respirator, and the patient in the second case could have given consent to the researcher to begin or continue the research when the patient becomes incompetent. Notice that specific authorization has been restricted to those cases where the instructions of the principal are specific on whether consent should be given or withheld. It does not cover those cases where the principal permits the agent to consent or refuse.

General Authorization with Instructions

In this situation the principal appoints an agent to either give permission or deny permission to any procedures at a time when the principal anticipates that he or she will not be competent, and the principal gives the agent instructions to follow in making the decision to give permission or not. The differences between this and specific authorization are: the instructions are general rather than specific, the agent may consent or refuse, and more than one procedure is placed within the authority of the agent. This kind of arrangement would be given legal recognition by the proposed Medical Treatment Decision Act in Michigan.[7] This bill authorizes one person to appoint another as his or her agent to accept or refuse medical treatment when the appointing person is incapable of accepting or refusing medical treatment. It also provides for the appointing person to write insturctions to the agent. Under the provisions of the bill a person may do either one without the other; if both are done, it would fit the model of general authorization with instructions. The language of the bill and the appointment form contained in it use the term "medical treatment." There was no express concern with medical research when the bill was drafted; whether the agent has the authority to accept participation in medical research for the principal would depend on the language of the instructions. The extent and clarity of the instructions will also be crucial in determining whether a particular treatment or research procedure is in compliance with the principal's wishes. This particular problem will be discussed at greater length in what follows.

General Authorization Without Instructions

This situation is the same as the one just described except that the principal does not provide instructions to the agent. The language used to appoint the agent will be essential to determine whether the agent is authorized to consent to treatment, research, or both. In order to maintain a difference between this model and general authorization with instructions, appointing an agent to make decisions of a given sort has to be distingushed from providing the agent with instructions for making decisions of that sort. A principal might appoint an agent to act on his or her behalf in all matters concerning the principal; this would be a carte blanche power of attorney. Or a principal might appoint an agent to make decisions only regarding the medical treatment of the principal. If the appointment of the agent were no more specific than that, it's doubtful that the agent would be within his or her authority to accept inclusion of the principal in a research protocol that held out no prospect of medical benefit to the principal. If, on the other hand, the principal specifically mentioned medical research as well as treatment, then it would not exceed the agent's authority to consent to participation of the principal. All of the above would be general authorization without instructions. If, in addition to specifying the sort of decisions that the agent may make, the principal instructs the agent on the kinds of medical treatment or research that the agent should consent to (or refuse), or if the principal specifies objectives and/or describes limitations on the research procedures, then it is general authorization with instructions. For example, a patient with Alzheimer's disease in its early stage could authorize another person to consent to research procedures, and instruct the agent to consent only to research relating to Alzheimer's that was non-invasive, and that had as an objective the development of drug therapy for memory loss and confusion.

Instructions Without Authorization

If a principal has not appointed anyone as his or her agent, decisions will have to be made for the principal by someone. If the principal has a close kin or friend who is available and willing to make decisions, then he or she becomes the agent. Such an agent has the authority to accept or refuse medical treatment, but this authority is conferred by law and custom, rather than by the express authorization of the principal, though the principal may expect that person to become his or her agent. This model provides for the agent to make instructions concerning medical decisions. The Living Will and the California Natural Death Act are examples. Both contain instructions on when life-sustaining procedures should be withdrawn, but neither clearly specifies anyone as the agent of the person who signs the document. The Living Will is addressed to "my family, my physician, my clergyman, any medical facility in whose care I happen to be and any individual who may become responsible for my health, welfare or affairs." The California statute titles the instructions as "Directive to Physician," and in the text there is the phrase, ". . . this directive shall be honored by my family and physician(s) . . . ." Since neither of these requires or encourages a person to appoint another particular person as agent, they are not one of the authorization models. Other examples of this model include situations where a principal devises instructions of his or her own, rather than using the Living Will or a statutory directive, but does not specify anyone as the agent. A principal might also orally address instructions to other persons generally; he or she might express to physicians, nurses, family, and friends what his or her wishes are in regard to medical treatment or research when he or she becomes incompetent, and in doing so, not designate anyone as agent. It is possible in the context in which instructions are written or spoken, that a clear intent is evidenced that some particular person is appointed as the agent of the principal; in that case, the conditions of general authorization with instructions are met.

Substitute Judgment

This model is like the previous one in that no agent is appointed; its second defining feature is that no instructions are given. Another feature of this model is drawn from the legal use of the notion of substitute judgment. If a close relative or friend of a principal is making a judgment about whether the principal should receive a particular medical treatment, the agent can, in some instances, determine that the principal would make a certain decision if he or she were competent to decide. The term "substitute" may not be the best one to use here, for it suggests that the agent is substituting his or her judgment for that of the principal in the sense that it is not the judgment that the principal would make. Whatever the aptness of the term, I restrict this model to those cases where the agent can say with confidence that the decision he or she makes is what the principal would want, i.e., the decision the principal would make if the principal were not presently incompetent. Substitute judgment is similar to general authorization with instructions when the agent in the latter can determine what course of action the instructions require in a given situation.

Deputy Judgment

This situation is like substitute judgment and instructions without authorization because the principal has not appointed an agent; it differs from substitute judgment in that the agent cannot say that the decision he or she will make is the decision the principal would make if the principal were competent. This may be because the agent does not know the principal; does not know the principal's values and attitudes well enough to be confident about what they imply in the situation at hand; or, though the agent is familiar with the principal's values and attitudes, there is nothing in them that points to any specific decision for the matter at hand. The agent will have to appeal to what is in the best interest of the principal, to what a reasonable person would want, to what he or she would want in a similar situation, or to some other standard that does not make appeal to any specific values known to be held by the principal.

It is important to emphasize that these six models are abstractions or ideal types. No claim is made that they are all in use, nor is it claimed that every actual situation falls neatly into one or the other of the six models. There can be uncertainty whether an agent was authorized by a principal or has to be determined by law and custom; there can be a wide range of specificity and generality in instructions, so it may be hard to say whether a given appointment is specific authorization or general authorization; there will also be borderline cases between substitute judgment and deputy judgment; finally, for the same principal and agent, some matters for decision may fit one model and others another. For example, a principal may authorize an agent to make the decisions about when the principal should be resuscitated in the event of cardiovascular arrest; this would be general authorization without instructions, but the same principal may have said nothing to the agent regarding medical research on the principal's underlying disease process. A decision to include the principal in research would have to be substitute judgment or deptuty judgment. Thus, the six models are not exclusive, and they may not be exhaustive, but they will prove useful for discussion of the main focus of this paper, viz., whether proxy consent can respect autonomy.

Autonomy and Proxy Consent

To simplify the discussion of the relationship between autonomy and proxy consent, I will assume that at the time the agent must accept or refuse participation of the principal in medical treatment or research, the principal is completely incompetent, i.e., he or she does not have any of the capacities required for making a decision regarding participation. Later I will consider what effects partial competence of the principal has on the relationship between autonomy and proxy consent.

It was shown that the concept of autonomy is not univocal, that autonomy may be predicated of actions, of a person's capacities, and of a person's relationship to external conditions. Further, it was shown that there are four aspects to the idea of autonomy as applied to actions: free action, authenticity, effective deliberation, and moral reflection. Six models of proxy consent were presented. The question now is: To what extent do the alternative models of proxy consent protect the autonomy of the principal in terms of the four aspects of autonomous action?

The first thing to examine is the autonomy of the act of authorization in the three models of proxy consent in which a principal appoints an agent, viz., specific authorization and general authorization with and without instructions. With respect to all three of them, it is plausible to inquire whether the act of authorization was a free action, whether it was authentic, and whether it resulted from effective deliberation. If an act of authorization has all of these characteristics then it is fully autonomous. The factors that would prevent an authorization from being fully autonomous are fairly obvious. If the principal were not fully competent at the time of the authorization, it may lack autonomy in one or more of its aspects. If the principal is being coerced, or what is more likely, manipulated or taken advantage of, then the authorization may not be a free action because it is not voluntary. Another possibility is that although the principal is not being manipulated and knows what he or she is doing, the principal is not capable of effectively deliberating on the alternatives and their consequences because of a cognitive deficit that makes concentration difficult. The authenticity of an authorization is likely to be an interesting feature with regard to the matter of who the principal chooses as an agent. If the principal chooses someone other than the person who has been closest in recent years, that can be reason to doubt that it is authentic, and lead to further inquiries about whether the principal is being manipulated. If there is no evidence of manipulation and it appears that the principal effectively deliberated about the choice of an agent, then it is a free action, and the result of effective deliberation, and should be respected as the autonomous action of the principal.

There is a feature of any act of autonomous authorization that is of great importance. When a principal authorizes another as an agent, then the principal makes the act of the agent an act of the principal. When the agent acts, that action is a free action of the principal in a secondary sense. This is what makes plausible, and sometimes expected or demanded, that one person accept responsibility for the acts of another. When one person voluntarily intends that another person act for him or her in a given situation, then the former person voluntarily intends the free action of the second person. In any case of proxy consent to medical treatment or research, if the agent has been authorized by the principal to accept or refuse medical treatment or research on the principal, then the acts of the agent are acts of the principal in this derived manner, and as such, have features of autonomy that are not present when the agent has not been authorized. I will refer to this feature of autonomy as free action in the secondary sense. We can now look at each of the six models of proxy consent to determine in what other ways they can respect the autonomy of the principal.

Specific Authorization

In this case the agent is merely a messenger of the principal, so whether the agent's consent or refusal respects the autonomy of the principal depends on whether the principal's act of authorization was autonomous. If the principal's act was autonomous in the sense of free action, authenticity, and effective deliberation, then the act of the agent to accept or refuse treatment or research, so long as it complies with the principal's specific instructions, is an act that respects the autonomy of the principal in these three senses. Concerning medical research, if it were possible to recruit a subject for a particular research protocol when the subject is competent, and it is expected that the research will be continued or be initiated when the subject becomes incompetent, then a prior consent of the subject, coupled with the proxy consent of an agent who has been specifically authorized by the principal, is as good a consent process as is first person consent—i.e., it can respect autonomy as fully as a first person consent. The most likely problem with this, assuming that the principal's authorization was autonomous, are situations where the conditions have changed from the time of the authorization to the time of the agent's action. If the expected condition of the patient and the circumstances of the research are not an explicit part of the authorization, but seem implicit in it, then if those conditions and circumstances change in some significant respect, there can be good ground for inquiring whether the principal's authorization and prior consent extend to the present situation. If this model of proxy consent is used, it is important to be as explicit as possible about the features and circumstances of the research when the principal's prior consent and specific authorization are obtained.

General Authorization with Instructions

Under this model the consent or refusal of the agent cannot be respectful of the autonomy of the principal simply in virtue of the autonomy of the act of authorization, as is the case with specific authorization, because the agent is more than a messenger for the principal. The action of the agent can be a free action of the principal in the secondary sense, as was discussed above, but cannot be a free action in the direct or primary sense.

In considering whether this model of proxy consent respects autonomy as authenticity and effective deliberation, the act of the agent to consent to or refuse the principal's involvement must be distinguished from the fact of the principal's involvement or lack of involvement in the medical treatment or research. Under this model the act of the agent to consent or refuse cannot be the act of the principal, and thus it can't be said that this is authentic or effective deliberation for the principal. Rather it is the fact that the principal is involved in the medical treatment or research that must be looked at to determine whether it is authentic for the principal, and whether it respects effective deliberation for the principal. The general instructions of the principal will be of obvious importance on the issues of authenticity and effective deliberation. If the instructions of the principal clearly reflect the known values and attitudes of the principal and if the action of the agent, whether consent or refusal, is consistent with these values and attitudes then the autonomy of the principal has been respected in the sense of autonomy as authenticity.

On the question of whether the agent's consent or refusal respects the autonomy of the principal as effective deliberation, there are three possibilities. Firstly, the instructions can be employed by the agent to deliberate on the alternatives open for the principal, and lead to accepting one of them as clearly implied by the instructions given the situation. This would respect the autonomy of the principal as effective deliberation, although in a secondary sense, i.e., the principal does not deliberate, the agent deliberates for the principal using the principal's instructions. If the instructions are sufficiently clear given the situation, and the alternatives and their consequences fairly straightforward given the instructions, then the effective deliberation of the agent can be the same deliberation the principal would have done if the principal could have.

Secondly, it is possible, given the instructions, the situation, and its alternatives, that consent and refusal are equally consistent with the instructions of the principal. In that case, the effective deliberation of the agent cannot be the effective deliberation of the principal in the secondary sense; the agent will have to appeal to something beyond the instructions of the principal to decide whether to accept or refuse participation of the principal.

Lastly, the instructions of the principal and the known values and attitudes of the principal may conflict; for example, the principal may have shown no interest in or commitment to medical research, and in fact the agent may know that the principal turned down opportunities in the past to participate in medical research. On the other hand, the instructions given the agent may authorize the agent to consent to the principal's participation in medical research of a certain sort under certain conditions. Whether the agent should follow the instructions or be consistent with the prior values and attitudes of the principal can be resolved by examining the autonomy of the act of authorization and of the preparation of the instructions. If those acts appear to be autonomous in the sense of free action and in the sense of effective deliberation, and if there is a plausible explanation for why the agent departed from previously expressed values in preparing the instructions (say that the principal was convinced in the process of his or her own recent illness that medical research is very important and that he or she wished to make a contribution to others), then the agent can follow the instructions and respect the principal's autonomy. If all this cannot be clearly worked out, then there is a serious dilemma for the agent, and no matter what decision is made, the autonomy of the principal will not be fully secured.

In summary, it is possible with this model of proxy consent to protect the autonomy of a principal in many respects; the agent's consent or refusal can be a free action of the principal in a secondary sense, it can respect the autonomy of the principal as authenticity and effective deliberation. When this is possible, and that will depend on the several sorts of factors discussed above, this form of proxy consent comes close to the model of specific authorization, which is very close to first person consent.

General Authorization Without Instructions

This model shares with all three authorization models the feature of free action in the secondary sense; the principal makes the action of the agent his by the act of authorization. It is possible for this model to respect the authenticity of the principal if the agent knows the values and attitudes of the principal and if they have clear implications regarding consent or refusal to a particular medical treatment or research protocol. Suppose that the principal had in the past frequently expressed an interest in medical research, had participated in more than one research protocol, and authorized the agent to consent to or refuse the principal's participation in medical research. The agent may then be quite confident that participation in a given project is implied by the principal's values and attitudes, and that the principal would consent if he or she could. If on the other hand, the principal has never in any way indicated his view of medical research, then the agent will not be able to consent with such confidence. Respect for autonomy as effective deliberation is not possible under the model of general authorization without instructions, for the agent does not have the instructions to apply to a given set of alternatives. In summary, under this model, proxy consent can respect the autonomy of the principal as free action in the secondary sense and as authenticity. The former is assured, the latter is not.

Instructions Without Authorization

This model cannot provide free action in the secondary sense, for the agent has not been appointed by the principal. There can be effective deliberation in the secondary sense, as is the case with general authorization with instructions. This model resembles the forementioned in another respect; if the agent knows the principal's values, attitudes, and life plans, and can confidently determine that they would lead the principal to accept or refuse a given medical treatment or participation in a particular reseach protocol, then it is possible for the agent's decision to be authentic for the principal, and it is possible for there to be a conflict between following the instructions and a decision based on the known values of the principal prior to the preparation of the instructions. The ways in which this possible conflict can be resolved were sketched above.

Substitute Judgment

In this model the agent is not authorized by the principal, but becomes the agent by law and custom. Thus, unlike the three authorization models, the action of the agent to accept or refuse participation of the principal cannot be free action of the principal in the secondary sense. Nor can there be effective deliberation in the secondary sense, for the agent does not have the principal's instructions to apply to the options. The only aspect of autonomy that is satisfied by substitute judgment is autonomy as authenticity. This follows from the definition; a necessary condition was that the agent knows the principal's values, attitudes, and life plans, and can confidently determine that they would lead the principal to accept or refuse a given medical treatment or participation in a particular research protocol.

Deputy Judgment

This final model obtains when none of the others do. There is no authorization of the agent by the principal, and the agent is not able to make a decision that is implied by the principal's values and attitudes. Thus, this model does not respect the autonomy of the principal in any of the four aspects of autonomy. The only way to assess the judgment of the agent is whether to and to what extent it advances or sacrifices the interest of the principal. When the principal has not authorized an agent, the agent is usually someone who knows the principal, and will know something about the principal's values and attitudes. So whether the judgment made by such an agent is a substitute judgment or a deputy judgment will depend on how well the agent knows the values of the principal, and whether they determine a specific decision with regard to the proposed medical treatment or research protocol. If the agent cannot say with confidence, "I know the principal's values and I know that this is what he would want," then the agent's judgment is deputy judgment, and none of the aspects of autonomy can be respected by the agent's decision.

Summary

The extent to which the six models of proxy consent satisfy the four aspects of autonomous action is compared in the following table. For each type of proxy consent, the listing of one of the aspects of autonomy means that it is possible to respect it; whether it is in fact respected in any given case will depend on the specific features of the situation.

1. Specific Authorization
   —Free action
   —Authenticity
   —Effective deliberation
2. General Authorization with Instructions
   —Free action (secondary)
   —Authenticity
   —Effective deliberation (secondary)
3. General Authorization without Instruction
   —Free action (secondary)
   —Authenticity
4. Instructions without Authorization
   —Authenticity
   —Effective deliberation (secondary)
5. Substitute Judgment
   —Authenticity
6. Deputy Judgment

Partial Competence of the Principal

All of the foregoing discussion of the extent to which the models of proxy consent realize or protect the autonomy of a principal proceeded under the assumption that at the time of the agent's decision, the principal was clearly incompetent. This is most likely to occur when the principal is unconscious or comatose. When the principal is partially competent or intermittently competent, respecting the autonomy of the principal by proxy consent becomes more complex, though not necessarily more difficult. The partial competence of a patient will affect the ways in which his or her own actions can be autonomous. A patient's cognitive impairment may be such that he or she cannot effectively deliberate—that is, the patient cannot concentrate on and compare several alternatives and their consequences. Yet, the patient may be able to understand a proposed procedure and give voluntary permission that the procedure be done, in which case the patient's decision is autonomous in the sense of free action and may be autonomous in the sense of authenticity. If what is proposed is medical treatment for the benefit of the patient, this degree of autonomy is sufficient to do the procedure. However, if the procedure is a research procedure that offers no benefit to the patient, then there is reluctance to accept the patient's consent as adequate. The reasoning behind this is obvious and persuasive; when we ask someone to make a sacrifice for others, we want to be sure that they at least have the opportunity, and hence the capacity, to effectively deliberate on the alternatives open to them.

When the patient does not have the capacity for some aspect of autonomy of action, an agent can supply it. Suppose the patient is capable of autonomous action in the sense of free action, but not in the sense of effective deliberation. If the patient has appointed an agent and provided the agent with general instructions for medical treatment and research, then when the principal gives his or her permission, and the agent gives his or her permission, the consent situation will be a combination of first person consent and proxy consent. It will be an upgrading of general authorization with instructions, for instead of free action in the secondary sense, there will be a free action of the principal. This is the kind of situation that the National Commission evisages in its reports on research on children and research on the mentally inform. Where a subject is not capable of consenting, the Commission speaks of the "assent" of the subject and the "permission" of a guardian. The discussion of the notion of assent shows that the Commission has in mind what I call free action.

> *The standard for "assent" requires that the subject know what procedures will be performed in the research, choose freely to undergo those procedures, communicates this choice unambiguously, and be aware that subjects may withdraw from participation. This standard for assent is intended to require a lesser degree of comprehension by the subject than would generally support informed consent. . . .*[8]

The permission of a guardian, or agent, allows the effective deliberation of the principal to be respected in the secondary sense if the general instructions of the principal clearly imply that the agent should give permission for the research procedure.

The assumptions in the above example of partial competence allow the autonomy of the principal to be increased over a situation where the principal is not competent to any degree. However, it won't always be so rosy. Suppose that the principal's partial competence is such that free action is possible, and that the principal assents to participation, but the agent firmly believes that on the instructions provided by the principal, he or she should refuse participation? Or it might be the other way around—the principal refuses and the agent believes he or she should consent. Should the partially autonomous action of the partially competent principal take precedence over the decision of the agent acting under the prior instructions of the principal? I have no reasoned answer to this difficult question for any but the easy cases, e.g., when the proposed procedure is for the benefit of the principal and the principal refuses it while the agent approves it.

One final, difficult situation occurs when the principal is intermittently competent, and at one point consents to a procedure and at a later point refuses it, and at one of the points appears more competent than at the other. If the decision concerns treatment for the benefit of the patient, there is less a problem than if the decision is whether to put the patient in a research protocol where the procedures are in part determined by the structure of the protocol and not entirely by what will benefit the patient. In the latter situation, respecting the autonomy of the patient is more likely if there is an agent who was explicitly authorized by the patient to consent to research of a certain sort, i.e., the model of specific authorization or the model of general authorization with instructions is satisfied. As the situation of the proxy consent moves away from these towards the model of deputy judgment, the autonomy of the patient is threatened and the research is less justifiable. What the general guidelines for this should be are beyond the scope of this paper.

Recommendations

The issue that motivated the structure and content of this paper is "Can proxy consent to research respect the autonomy of an incompetent research subject?" The answer is a qualified "Yes." If the concept of autonomy is fully analyzed and the various ways proxy consent can be given are sorted out, it is seen that in some situations proxy consent can respect much of an incompetent subject's autonomy.

What policy recommendations can be made on the basis of the relationships between autonomy and proxy consent? In order to answer this question, attention must be given to how much autonomy can be protected by the various models of proxy consent and to the practical problems of utilizing the models of proxy consent. From the first perspective, the three authorization models of proxy consent have a distinct advantage over the remaining three. Whenever a principal appoints an agent and gives the agent authority to make decisions concerning a particular matter affecting the principal, the principal has made the agent's actions those of the principal. This respects the principal's autonomy as free action, even though it is in a secondary sense. This feature is more important than the possibility of respecting authenticity or effective deliberation. The reason for this is that even when the agent knows the principal's attitudes and values or has the principal's general instruction, they underdetermine a decision to accept or refuse participation in research. Applying values to concrete situations is not a matter of simple deduction, because to hold a particular value is to have a general disposition, and because concrete situations do not come labeled with the values they will realize and thwart. Further, everyone holds values that can lead to opposite decisions in a given situation. Thus, the process of applying values, whether one's own or another's, is also a process of determining the scope and relative importance of values. This means that when one person is applying another person's values or instructions, there is always a strong possibility that if the first person applied the values, a different decision would have been made. So there is always reason to doubt that authenticity or effective deliberation is really being respected as it would be if the principal were making the decision. If the agent applying the values is someone chosen by the principal to do so, then uncertainty about the application is lessened in importance, though not removed, by the important fact that the principal has entrusted the agent with the task of applying the principal's values or instructions, and has made the agent's decision his by the act of authorization. Thus the three authorization models are generally preferable to the three without authorization from the point of view of respecting the autonomy of the principal.

Of the three models of proxy consent without authorization, deputy judgment by definition does not respect the autonomy of the principal. Instructions without authorization and substitute judgment can respect autonomy, but they have the problem of underdetermination of decisions, as discussed above, without the advantages of authorization. Though it is possible for autonomy as authenticity or effective deliberation to be respected by these two models, the likelihood of that happening, or for anyone to have any confidence that it has happened, are slim. Thus, a policy on proxy consent to research should not encourage the use of any of the non-authorization models of proxy consent.

Consideration will now be given to the practical aspects of utilizing the three authorization models of proxy consent. To obtain the greatest amount of respect for autonomy, policy could require or place priority on specific authorization. The practical limitations of this are obvious. It's not often that a person could be informed while competent of the specific research protocols that the person will be suitable for when and if that person becomes incompetent. For a genetically determined progressive disease, e.g. Huntington's chorea, it is possible to obtain the prior consent of a person to research on the disease, and to have the person appoint an agent to consent to research. But unless the researcher can specify the exact nature of the research protocol at the time of obtaining the authorization, this would not be specific authorization. The model of specific authorization could be used in a situation where a patient has a disease process that has a significant probability of causing the patient to become incompetent within a relatively short time, and there are one or more protocols for which the patient is suitable and that would continue, or be initiated, after the patient becomes incompetent. Consent could be obtained for whatever will be done while the patient is competent, and the patient could appoint an agent with specific authorization to consent to continuing or initiating research procedures after the patient becomes incompetent. The amount and kinds of research where this is possible is probably very limited, but since specific authorization does protect autonomy to the greatest extent, it should be utilized whenever possible.

General authorization with and without instructions has a much wider possibility of use, since it does not require at the time of authorization that the principal know what specific procedures are proposed. There are two ways in which general authorization could be used. On the one hand the general approach of the proposed Michigan Medical Treatment Decision Act could be extended to include research procedures. The idea is that healthy individuals would appoint another to consent to or refuse medical treatment and research on their behalf when they become incompetent. Although this is worth making possible by legislation, it is not likely that enough persons would do this so that a researcher could identify a sufficient number of incompetent patients at the time of initiating a particular research protocol who had previously made a general authorization for research. An approach that is more likely to work is for researchers to identify persons who have conditions that make them potential subjects in a general area of research. This would be done for patients who are known to have particular diseases and who are still sufficiently competent to appoint an agent. Since the authorization is general, it would not be necessary to know the specific research protocols in which the patient would be included. In research on Alzheimer's disease and related types of senile dementia, patients would have to be approached during the early stages of the disease when, though they may suffer some memory deficit and occassional disorientation, nonetheless they have sufficient capacity to make a fully autonomous decision to participate in research, to appoint an agent, and to give the agent instructions. If this is done, research can be initiated while the patient is competent and can be continued or initiated when the patient becomes incompetent, and the patient's autonomy will not be seriously compromised.

In conclusion, it is possible to respect autonomy with proxy consent, based on a full analysis of those two concepts. The key to linking them lies in the process of one person authorizing another to act in the first person's behalf.

Notes and References

[1]45 CFR Part 46, 46.111, (a)(4).

[2]National Commission for the Protection of Human Subjects of Biomedical and Behavioral Research, *Report and Recommendations: Research Involving Those Institutionalized as Mentally Infirm,* DHEW Publication No. (OS) 78–0006. pp. 1–22. *Report and Recommendations: Research Involving Children,* DHEW Publication No. (OS) 77–004. pp. 1–20.

[3]Paul Ramsey, *The Patient as Person* New Haven: Yale University Press, 1970, pp. 11–19.

[4]Richard McCormick, "Proxy Consent in the Experimentation Situation," *Perspectives in Biology and Medicine* 18; Autumn 1974, 2–20.

[5]K. Steel and R. Feldman, "Diagnosing Dimentia and Its Treatable Causes," *Geriatrics* 34; March 1979, 79–88.

[6]Further explanation in B. Miller, "Autonomy and the Refusal of Life-Saving Treatment" *Hastings Center Report* 11; August 1981, 22–28.

[7]B. Miller, "The Michigan Medical Treatment Decision Act," in C. Wong and J. Swazey, *Dilemmas of Dying: Procedures and Policies for Decisions Not to Treat* (Boston: G. K. Hall, 1981), pp. 161–168.

[8]*Research Involving Those Institutionalized as Mentally Infirm,* p. 9.

Derived Consent, Proxy Consent

Legal Issues

Lance Tibbles

Common Law Consent Requirements

It has been established in recent years in the United States that "informed consent" is necessary before a medical procedure may be performed on an individual. This consent may be an assent by a competent person, an incompetent person's guardian, or by someone closely identified with a person not thought capable of giving informed consent.[1] A competent adult generally cannot be subjected to medical treatment without his "informed consent."[2] But a child or an adult who has been judicially declared incompetent to care for himself is thought to lack the capacity to understand the nature and purpose of the medical procedure, and to weigh the risks and benefits involved. Therefore, individuals who lack legal competency cannot make legally binding decisions about their medical care. Other parties—the parents of a child and a court appointed guardian for an adjudicated incompetent adult—may assume the decision-making role to protect the incompetent from his own lack of knowledge and from coercive methods used to obtain his consent.[3]

There are theoretical difficulties when dealing with a person for whom some doubt exists about capacity, but for whom there has been no legal guardian appointed.

> *. . . "consent" for a procedure will have to come from someone besides the patient whenever the latter is legally incompetent. If the patient is the primary beneficiary of the procedure, this presents some problems but not particularly taxing ones. For example, when a patient is admitted to an emergency room, his condition can be such (due to*

> *cardiac arrest or acute renal failure) that he is unable to participate in the deliberations about his treatment, and the decision must then be made instead by a member of his family; this was the situation with the first heart transplant in man, in which the recipient's sister gave permission for the operation. As a theoretical matter, difficulties certainly rise when any person is given power over another, and these may be exacerbated rather than reduced (as is assumed by the law) when the persons involved are members of the same family. Yet there are reasons of sentiment, convenience, and even good sense for this allocation of authority, and it is a practice which is so well known in society at large that any individual who finds the prospect particularly odious has ample warning to make other arrangements better suited to protecting his own ends or interests.*[4]

The proxy—the parent, the legal guardian, or other proxy decision-makers—may act only for the incompetent's benefit or welfare. In addition, the courts acting under their *parens patriae* power—the state's duty to care for those who are unable to care for themselves—may intervene when there is a question as to whether the proxy decision-maker is acting in his ward's best interest. Thus, in potential conflicts of interest between the guardian and the ward, such as when unusual or experimental high-risk medical procedures are involved, courts will often review proxy consent to assure that it serves its intended purpose—the protection and promotion of the ward's personal autonomy.[5] We act in the best interests of the ward by protecting and promoting his personal autonomy.

At this point, it is important to digress for a moment to distinguish between (1) medical procedures that have no purpose other than the well-being of the individual patient and (2) medical procedures that also serve other interests or that are carried out solely to obtain information of use to others, without intent to treat any illness of the individual. Professor Charles Fried gives the following analysis of experimentation in medical procedures:

> *At the outset we must distinguish between therapeutic and nontherapeutic experimentation. Experimentation is clearly nontherapeutic when it is carried out on a person solely to obtain information of use to others, and in no way to treat some illness that the experimental subject might have. Experimentation is therapeutic when a therapy is tried with the sole view of determining the best way of treating that patient. . . . Much research is mainly therapeutic, in the sense that the patients' interests are foremost, but nevertheless things*

> *may be done which are not dictated solely by the need to treat that patient: tests may be continued even after all the information needed to determine the best treatment of the particular patient have already been completed; or substances may be injected for a period or in doses not strictly necessary for the cure of that patient, but with the motive of developing information of use to others . . . it must be recognized that persons who become research subjects in nontherapeutic experimentation may often be the beneficiaries of a degree of medical attention which they might not otherwise enjoy, and which thus redounds to their benefit.*[6]

Professor Fried then sets out three classes of medical procedures.

Therapeutic

> *Legal decisions and commentators have always stated that a practitioner is only justified in using "accepted remedies," unless his patient specifically consents to the use of an "experimental" remedy. . . . General principles require the consent of the patient to any therapy, usual or unusual. It is just that as the therapy moves away from the standard and the accepted, the need for explicit consent, full disclosure of risks and alternatives, becomes more acute, and more likely to pose an issue. . . .*
>
> *The obligation to advise the patient of alternative therapies does not extend to all the hypothetical, untried or experimental remedies that various researchers are in the process of developing. Where, however, the therapy used is itself experimental, then this fact and the existence of either alternatives or professional doubts become material facts, which like all material facts should be disclosed. Beyond this, where the experimentation is truly and exclusively therapeutic there are no particular legal constraints that do not apply to the practice of medicine generally.*[7]

Mixed Therapeutic and Nontherapeutic

> *The kind of medical experimentation which causes the greatest legal and ethical perplexities is what might be called mixed therapeutic and non-therapeutic experimentation: The patient is indeed being treated for a particular illness, and a serious effort is being made to cure him. The systems of treatment, however, are not chosen solely with the view to curing the particular patient of his particular ills. Rather, the treatment takes place in the context of an experiment or a research program to test new procedures, or to compare the efficacy of various established procedures. Nor is it the case that this research*

purpose is limited to carefully reporting the results of treatments in particular cases. Rather, therapies are tried, continued or varied, and patients are assigned to treatment categories partially in response to the needs of the research design, i.e., not exclusively by considering the particular patient's needs at the particular time.[8]

Nontherapeutic

No special doctrines apply to nontherapeutic experimentation. Indeed, to the extent that the experimentation is nontherapeutic, the fact that it is being carried out by doctors should be entirely irrelevant. The usual privileges under which doctors work, and the usual special doctrines according to which the liabilities of doctors are judged should not be applicable, since they proceed from the premise that the doctor must be given considerable latitude as he works in the presumed interests of his patient. But that is not the case in nontherapeutic research. The doctor confronts his subject simply as a scientist.[9]

Statutory Consent Requirements

Two types of statutes providing for proxy consent to therapeutic procedures are common in the United States. First, there are statutes that set out the situations—including proxy consent—that satisfy the requirements of informed consent in the general medical setting. Most of these statutes contain rather broad, general statements describing the situations in which proxy consent must be obtained for an adult who might be incapable of giving an informed consent, e.g., the patient is "of unsound mind,"[10] "competence, infancy, or . . . the patient is under the influence of alcohol, hallucinogens, or drugs, lacks legal capacity to consent";[11] a person other than a "person of ordinary intelligence and awareness sufficient for him or her generally to comprehend the need for, the nature of and the significant risks ordinarily inherent in any contemplated hospital medical, dental, or surgical care, treatment or procedure";[12] and there is a person "standing in loco parentis . . . for his ward or other charge under disability."[13] In these statutes the identity of the authorized proxy decision-maker is also described broadly, e.g., "patient's spouse, parent, guardian, nearest relative or other person authorized to give consent for the patient,"[14] "a person who has legal authority to consent on behalf of such patient in such circumstances,"[15] "parent, spouse or guardian. . . . [and if none of these are readily available] any competent relative . . . or . . . any other competent individual."[16]

The second type of statutes limit the performance in state mental hospitals of certain medical procedures, usually surgery, without consent, including proxy consent, of a patient. In these statutes the identity of the proxy is also broad, e.g., "a responsible member of his family or of a guardian,"[17] "parent, guardian, spouse, or adult next of kin,"[18] "spouse, guardian, either parent, or oldest adult child."[19] Several of these statutes also include broad authority for the head of the hospital or chief executive officer of the institution to give the required consent, e.g., the condition "is of an extremely critical nature,"[20] and "the treatment to be performed is essential and beneficial to the general health and welfare of such patient or inmate, or will improve his opportunity for recovery or prolong or save his life."[21]

This vagueness, in both common law and statutes—both for the conditions that will allow proxy consent to therapy and the identity of the proxy agent—indicates that society has not resolved the difficult issues involved in authorizing proxy decision-making for serious medical procedures. This ambivilance makes it difficult to anticipate how society will ultimately respond to the difficult issues involved in proxy consent to nontherapeutic procedures in cases of Senile Dementia of the Alzheimer's Type (SDAT).

Recently, a few states have enacted statutes regulating the use of highly intrusive organic therapies on patients in state mental institutions. The most frequently regulated procedures are electroconvulsive therapy (ECT) and psychosurgery.[22] These statutes restrict access to these therapies by requiring that the physician explain to the patient and a responsible relative or guardian the procedures to be used in their treatment; the benefits, risks, side effects, and their degree of uncertainty; the nature and seriousness of the patient's disorder; and the patient's right to withdraw consent at any time. Most importantly, a review committee must agree that the procedure is indicated, and that the patient has the capacity to consent. California allows proxy consent for electroconvulsive therapy, but not for psychosurgery.[23]

Right to Refuse Consent to Life-Prolonging Procedures

One line of proxy consent has been developing more quickly, however. There are now a number of cases generally recognizing the right of an incompetent patient to refuse life-prolonging medical procedures.[24] The analysis has varied and there still is disagreement whether the advance consent of a court is required before life-prolonging measures can be either withheld or withdrawn. However, the courts generally agree that unless there is a compelling state interest that will be served by continuing the life-prolonging treatment, the constitutional right of privacy protects an incompetent terminally ill person's right to refuse life-prolonging treatment even though the right is exercised by family members or other close friends. In addition, 22 states and the District of Columbia have enacted statutes specifying a procedure by which a person can prepare a document while he is competent (generally called an advance directive or a "living will"), stating that if he becomes simultaneously incompetent and terminally ill, he wishes to have withheld or withdrawn medical procedures that serve only to prolong his life.[25]

Nontherapeutic Procedures and No Duty to Confer a Benefit

In the case of ordinary therapy, at least in theory, the physician's only concern is the patient's well-being. But when we move to cases of untried therapeutic, mixed therapeutic and nontherapeutic procedures, not only are there more uncertainties and greater risks, but the physician who contemplates the procedure is motivated, at least in part, by a search for scientific knowledge. The patient–physician relationship is altered by the broadened objectives of the physician–researcher, who may no longer be sufficiently disinterested to be an objective participant. Thus, it is more likely that in this situation the law will be more strict and more protective of the subject's rights in the informed consent process.[26]

In our society, there is no general legal duty to rescue another person who is at risk of injury or disease. There are some exceptions. Exceptions exist when there is a pre-existing duty of care, such as the duties of a parent to his child and when the putative rescuer is responsible for placing the other person in a position of peril. We can also compel persons to submit to vaccinations, and persons with contagious diseases can be subjected to compelled quarantine. However, this intervention is intended primarily to protect third parties or the public generally, and is upheld upon that basis. But we have not gone further and compelled a person to confer a benefit against his will in cases other than those where some unique condition of that patient places the public at special risk. This is true even though the putative rescuer will not be placed at risk and the victim will surely be harmed without aid. Neither case law nor statutes require a person to donate an organ, or bone marrow, or a rare blood type, despite the fact that the risks are small, and a known, identified person will die unless he receives a transplant. This is true even though this refusal to donate would be regarded as immoral under applicable moral theory.

A recent case involving a possible bone marrow transplant offers a paradigm example of our refusal to require a person to confer a benefit upon another. In *McFall v. Shimp,*[27] McFall suffered from the rare bone marrow disease of aplastic anemia. His chances of survival without a bone marrow transplant were virtually nonexistent. Although the bone marrow donor experiences some pain and discomfort, the risk is minimal. Before an individual can become a candidate for donating bone marrow, two tissue compatibility tests must be performed. Shimp, a cousin, voluntarily underwent the first test that established that there was an excellent chance that Shimp would be a suitable donor and that no other family member would be appropriate. Shimp refused to submit to the second test. McFall brought suit to compel his cousin Shimp to submit to the second test, and, should the results indicate compatibility, eventually submit to a transplant. The Pennsylvania trial court squarely faced the following issue: in order to save the life of one of its members by the only means available, may society infringe upon another person's right to bodily security? The court answered the question in the negative and denied McFall's request. McFall died of his disease two weeks later. The court reasoned as follows:

> *The common law has consistently held to a rule which provides that one human being is under no legal compulsion to give aid or to take action to save that human being or to rescue. A great deal has been written regarding this rule which, on the surface, appears to be*

> *revolting in a moral sense. Introspection, however, will demonstrate that the rule is founded upon the very essence of our free society. . . . Our society, contrary to many others, has as its first principle, the respect for the individual, and that society and government exist to protect the individual from being invaded and hurt by another. . . . For our law to compel the Defendant [Shimp] to submit to an intrusion of his body would change every concept and principle upon which our society is founded. To do so would defeat the sanctity of the individual and would impose a rule which would know no limits and one could not imagine where the line would be drawn. . . .*

In cases of organ donation or human experimentation, the patient/subject who refuses to submit to a procedure does not constitute a danger to others, but merely refuses to confer a benefit. Competent adults may give their informed consent to donate organs or to be research subjects. But, failure of the physician/researcher to abide by the requirements of fully informed consent or failure to respect a competent adult's refusal to participate in nontherapeutic procedures is not justified.[28]

Because nontherapeutic procedures, by definition, are not for the subject's benefit, it has been stated that no proxy consent for incompetent individuals is theoretically permissible.[29] Whenever consent is sought for a procedure that is not solely for the patient's benefit, i.e., mixed therapeutic and nontherapeutic, it is generally assumed that a guardian lacks the authority to give consent.[30] However, there are some circumstances in which nontherapeutic procedures are allowed to be performed on incompetents. Organ transplantation and sterilization are the two most common cases.[31]

In the organ donation cases involving a child or a mentally retarded adult as donor for a sibling, the courts frequently find that the donor is really benefited by the procedure because of the impact of the death of the sibling upon the incompetent. The factors involved in this type of situation, including the fact that a specific life will be saved in exchange for the imposition of a minimal risk on the incompetent donor, make these cases rather unique. In addition, a concept of family unity in making intrafamily decisions of this type is also involved, making this line of cases inapplicable to other instances of nontherapeutic procedures on incompetents.[32]

The sterilization cases also proceed along the lines of the best interests of the incompetent. However, a state interest in preventing the birth of mentally retarded infants for whom the state or other individuals will have to care for is often used to reinforce arguments based on the incompetent's best interests. But, sterilization cases are limited to circumstances in which a valid state interest is thought to outweigh the rights of the individual so as to justify use of the police power in this manner. This exception is also rather narrow.[33]

Annas, Glanz, and Katz, while discussing human experimentation with institutionalized mentally infirm individuals, reach the following conclusions:

> *It is difficult at this time to make a definitive rule about non-therapeutic experimentation based on the law. In general, competent patients may consent to participation.*
>
> *. . . When the need for the information is great, and the risk to the individual participant minimal, this type of research should probably be permitted with incompetent patients as well, assuming that proxy consent has been obtained. Examples of procedures included in this category are the taking of blood and the collection of urine specimens. However, the refusal of an incompetent person to involvement in the experiment should be binding, regardless of either his reasons for the decision or the wishes of the patient's guardian . . . Non-therapeutic research is justified only when the condition under investigation is related to mental disability and the information sought cannot be obtained from noninstitutionalized subjects.*[34]

Professor Alexander Capron structures the issue as follows:

> *[There is] a formulation which builds on the present system in allowing permission for nonbeneficial interventions to be given on behalf of incompetents by someone, but which demands a clearer recognition of the competing interests. Only by acknowledging that certain interests are at war and then by attempting to balance them in each case can progress be made toward a satisfactory resolution of this problem. On the one hand, there is the interest a person has in being protected from abuse and exploitation should he become unable to protect himself. On the other hand, each person has an interest in not having the right to choose taken away awhile he is still able to make choices which reflect his own view of his goals and values.*
>
> *In the case of the adult competent to make his own choices, the former interest is protected through the latter. For someone who is legally incompetent, however, the latter interest does not operate, and attention is focused entirely upon the former.*[35]

Professor Jay Katz and Alexander Capron later complete the analysis:

> *If we are left with the unresolved question of when, for what purposes, and by whom permission may be given in lieu of personal informed consent, it would at least be useful to note the small advance which could be achieved by ceasing to call such substitute permission "consent." The reasons for relying on consent . . . relate back to the respect owing the individual and his right to autonomy and integrity. Wherever possible, the law tries not to second-guess decisions which a person makes for himself, as is reflected, for example, in the common law courts' unwillingness to evaluate the worth of the consideration which the parties to a contract accept as binding them to perform their agreement. Similarly, an agreement reached between doctor and patient following the disclosure and discussion contemplated by our informed consent model should be immune from attack in subsequent legal proceedings. Just as the law must be concerned that as a general principle patient–subjects are adequately protected by the methods that are employed to gain consent, so too it must give meaning to the process of making a choice by holding the parties to the burdens and costs inherent in their choices. Where the permission is given on behalf of someone else, however, these considerations do not attach. Accordingly, it would not only be acceptable but also advisable for such grants of permission to be subject to review as to their competency and motivation, so as to screen out those which are made unwisely or maliciously.*[36]

Analogy to Those Institutionalized as Mentally Infirm

For purposes of analysis of proxy consent issues there are two types of individuals to whom SDAT individuals may be analogized. One could use an analogy of "children." However, one important factor in analyzing proxy consent in children, especially for nontherapeutic procedures, is that the child has never been legally competent and his mental faculties are, for the most part, developing and improving with age. The child has no history of competent decisions for us to look to for guidance in determining what a competent adult version of him would do in a given situation. However, the analogy of "those institutionalized as mentally infirm" may prove to be a closer fit. Here we are dealing mostly with adults, all of whom suffer from some degree of mental incapacity. Some, but not all, of the advance stage SDAT patients may be institutionalized. Thus, SDAT individuals will share more characteristics with the institutionalized mentally infirm than with children. The National

Commission for the Protection of Human Subjects of Biomedical and Behavioral Research has issued reports and recommendations for both children and those institutionalized as mentally infirm.[37] Although both provide helpful analyses for us, the Commission's recommendations for the institutionalized mentally infirm offer a viable framework for beginning our analysis.

Three recommendations in the National Commission's Report and Recommendations on Research Involving Those Institutionalized as Mentally Infirm are worth setting out in some detail.

> *Recommendation (2) Research that does not present more than minimal risk to subjects who are institutionalized as mentally infirm may be conducted or supported provided an Institutional Review Board has determined that:*
>
> . . .
>
> *(B) Adequate provisions are made to assure that no subject will participate in the research unless:*
>
> *(I) The subject consents to participation;*
>
> *(II) If the subject is incapable of consenting, and research is relevant to the subject's condition and the subject assents or does not object to participation; or*
>
> *(III) If the subject objects to participation, the research includes an intervention that holds out the prospect of direct benefit for the individual subject and the subject's participation is specifically authorized by a court of competent jurisdiction.*[38]

The Commission's recommendation is based upon the research being relevant to the subject's condition and presenting no more than minimal risk. The Commission defines "minimal risk" as "the risk (probability and magnitude of physical or psychological harm or discomfort) that is normally encountered in the daily lives or in the routine medical or psychological examination of normal persons. Thus, for subjects who are institutionalized as mentally infirm, routine examination procedures present no more than minimal risk if the likely impact of such procedures on them is similar to what would be experienced by normal persons undergoing the procedures." The Commission uses "assent" as the applicable standard for the subject's agreement to participate when the subject is incapable of giving informed consent but other conditions are satisfied—the research is relevant to the subject's condition and there is no more than minimal risk. Here, if the subject is incapable even of assenting, the Commission says that absence of objection should be sufficient to permit participation. But if the subject objects to participation, it may not be authorized except by court order. As will be set out in more detail later, it is this writer's position that if any subject objects to any mixed therapeutic–nontherapeutic or nontherapeutic procedures, he or she should not be compelled to participate, even by court order, unless while competent that subject has appointed a durable power of attorney or executed a "Ulysses contract."

> *Recommendation (3) Research in which more than minimal risk to subjects who are institutionalized as mentally infirm is presented by an intervention that holds out the prospect of direct benefit for the individual subjects, or by a monitoring procedure required for the well-being of the subjects, may be conducted or supported provided an Institutional Review Board has determined that:*
>
> . . .
>
> *(B) Such risk is justified by the anticipated benefit to the subjects;*
> *(C) The relation of such risk to anticipated benefit to subjects is at least as favorable as that presented by alternative approaches;*
> *(D) Adequate provisions are made to assure that no adult subject will participate in the research unless:*
> *(I) The subject consents to participation;*
> *(II) If the subject is incapable of consenting, the subject assents to participation (if there has been an adjudication of incompetency, the permission of a guardian may also be required by state law);*
> *(III) If the subject is incapable of assenting, a guardian of the person gives permission (if a guardian of the person has not been appointed, such appointment should be requested at a court of competent jurisdiction) or the subject's participation is specifically authorized by a court of competent jurisdiction, or*
> *(IV) If the subject objects to participation, the intervention holding out the prospect of direct benefit for the subject is available only in the context of the research and the subject's participation is specifically authorized by a court of competent jurisdiction;. . . .*[39]

The Commission states that more than minimal risk is permitted only by an intervention that holds out the prospect of direct benefit to the individual subjects (therapeutic) or by a monitoring procedure necessary to maintain the well-being of those subjects (mixed therapeutic–nontherapeutic), e.g., when all available treatments for a serious condition may have been tried without success, and the remaining option is a new intervention presently under investigation. By "direct" the Commission says that the possibility of benefit to the subject must be fairly immediate.

Absence of objection is not sufficient grounds to proceed with research presenting more than minimal risk. However, the Commission provides for three ways in which a subject who is incapable of either giving informed consent or of assenting can participate in such research. First, the Commission allows a legally appointed guardian to give permission for the ward's participation. Second, if there is no legally appointed guardian, a court can be requested to appoint a guardian ad litem who can give permission for the ward's participation. Third, a court can be requested to specifically authorize such participation.

If the intervention expected to provide direct benefit to the subject is available only in the context of the research, the Commission would allow a court to override the objection of an adult subject.

> *Recommendation (4) Research in which more than minimal risk to subjects who are institutionalized as mentally infirm is presented by an intervention that does not hold out the prospect of direct benefit for the individual subjects, or by a monitoring procedure that is not required for the well-being of the subjects, may be conducted or supported provided an Institutional Review Board has determined that:*
>
> . . .
>
> *(B) Such risk represents a minor increase over minimal risk;*
>
> *(C) The anticipated knowledge (I) is of vital importance for the understanding or amelioration of the type of disorder or condition of the subjects, or (II) may reasonably be expected to benefit the subjects in the future;*
>
> *(D) Adequate provisions are made to assure that no adult subject will participate in the research unless:*
>
> *(I) The subject consents to participation;*
>
> *(II) If the subject is incapable of consenting, the subject assents to participation (if there has been an adjudication of incompetency, the permission of a guardian may also be required by state law); or*
>
> *(III) If the subject is incapable of assenting, a guardian of the person gives permission (if a guardian of the person has not been appointed, such appointment should be requested at a court of competent jurisdiction."*
>
> *The subject should not be involved in research over his or her objection.*[40]

The Commission's view is that individuals who are institutionalized as mentally infirm may participate in research presenting a minor increment of risk above minimal, even if there is no expectation that they will derive direct (fairly immediate) benefit from such participation (i.e., it is a nontherapeutic procedure), if either of two conditions are present. First, there is a good reason to believe the research will yield information of vital importance for the understanding of the condition for which the subjects have been institutionalized. This would include a case where the expectation may be only the development of better methods of diagnosis or prevention, so that others who are at risk for the disorder or a future generation of persons suffering from the disorder will be the beneficiaries of the research. Or, second, there is a possibility of remote benefit to the subjects, such as the eventual development of better treatment for their condition.

Application of the National Commission's Recommendations to SDAT Individuals

We may now take the Commission's recommendations for those institutionalized as mentally infirm and apply them, where applicable, to SDAT research involving incompetent SDAT individuals. We will not be concerned here with those cases where the proposed subject is competent and gives his informed consent to participate. Application of the Commission's recommendations to research upon incompetent SDAT individuals who are incapable of giving informed consent would permit research relevant to the subject's condition in the following cases:

1. Research that does not involve more than a minimal risk and
   a. the subject assents or does not object to participation; or
   b. the subject objects to participation, but a court specifically authorizes the subject's participation for either (1) an intervention that holds out the prospect of direct benefit to the subject, or (2) a monitoring procedure required for the subject's well-being.

2. Research involving more than minimal risk that either holds out the prospect of direct benefit to the subject or is a monitoring procedure required for the subject's well-being, and the risk is justified by the anticipated benefit to the subject and the relation of such risk to the anticipated benefit to subjects is at least as favorable as that presented by alternative approaches, *and*
   a. the subject assents to participation and the court appointed guardian, if any, also gives permission; or
   b. although the subject is incapable of assenting, a court appointed guardian of the person, if any, gives permission, or a court specifically authorizes the subject's participation; or
   c. although the subject objects to participation in intervention holding out the prospect of direct benefit available only in the context of the research, a court specifically authorizes the subject's participation.
3. Research that involves only a minor increase over minimal risk and does not hold out either the prospect of direct benefit to the subject or a monitoring procedure that is not required for the subject's well-being, but the anticipated knowledge either is of vital importance for the understanding or amelioration of SDAT, or may be reasonably expected to benefit future SDAT victims and
   a. the subject assents to participation and the court appointed guardian, if any, also gives permission; or
   b. although the subject is incapable of assenting a court appointed guardian of the person gives permission. The subject should not be used in this type of research if he objects to participating.

Subject Advocate, Durable Power of Attorney, and Ulysses Contract

The recently proposed[41] and even more recently tried[42] "subject advocate" for community-based medical research facilities may also play a role in research involving SDAT persons. Professor John Robertson has suggested that in institutions conducting a large volume of research a "subject advocate" position on the staff of the institutional review board may be appropriate to "explain rights, clarify risks and benefits, assist in decision making, observe or monitor research, and receive complaints."[43] In SDAT research the advocate could continuously assess each subject's comprehension, competency, voluntariness, and awareness of their part in the research. The advocate could be the first one to ascertain that the subject, whether competent or incompetent, does not wish to continue participating in the research for whatever reason—rational or irrational. The advocate could also facilitate information between the subject and his attorney in fact and/or guardian, between the subject and the institutional review board, and between the subject and the researchers. Perhaps the advocate's most important role would be to make sure that the subject's attorney in fact and/or guardian is continuously and immediately informed of any changes in the subject's condition and willingness to continue participating in the research.

The relatively new legal device of "durable power of attorney" may play an important role in decision making for incompetent individuals. It was not considered by the National Commission, but was advocated by the President's Commission for the Study of Ethical Problems in Medicine and Biomedical and Behavioral Research for decisions on withholding or withdrawing life-prolonging medical treatment.[44] A power of attorney is a written instrument by which one person, as principal, appoints another as his agent and confers upon him the authority to perform certain specified acts or kinds of acts on behalf of the principal.[45] Any person having the capacity to appoint an agent may confer upon another the power to act as his attorney in fact. A power of attorney can be terminated by the act or agreement of the parties or by operation of law. The parties can agree that the power will be terminated at the expiration of a specified or reasonable time or upon the accomplishment of the purpose of the power. In addition, the principal can revoke and the agent can renounce the power at any time, although the reneging party may be liable for damages caused to the other party in so doing.

The power of attorney is usually terminated by operation of law upon the death of either the principal or the attorney. Generally the attorney's authority is terminated or suspended upon the principal's loss of capacity to become a party to the transaction.[46] Thus, the traditional power of attorney is of little usefulness in the cases we are concerned with, because the attorney's authority to make decisions about medical interventions will terminate when the principal becomes incapable of making the decision himself.

The Uniform Probate Code, which has been enacted in many states, and the comparable new Uniform Durable Power of Attorney Act, which has been enacted in nine states, permit a competent individual to execute a power of attorney that will become effective or remain effective in the event he should later become disabled.[47] If the court subsequently appoints a guardian for the individual, the attorney is responsible to the guardian as well as the principal and the guardian has the same power to amend or revoke the attorney's powers that the principal would have had if he had not become incapacitated. Under these provisions, the written power of attorney can provide that "this power of attorney shall not be affected by subsequent disability or incapacity of the principal." Or it can use similar words showing the principal's intent that the agent is to have the power to act, notwithstanding the principal's subsequent disability or incompetence. Acting under the authority of this type of provision, a competent individual can appoint an attorney to make medical care decisions for him, and provide either that the authority is to continue notwithstanding his becoming incompetent or that the authority is to begin when he becomes incompetent.

The attorney has a duty of good faith and loyalty to the principal. Although the issue has yet to be raised in the courts, there is reason to believe that this duty of good faith would not ipso facto be violated by the attorney's consenting to the principal's becoming a subject of nontherapeutic experimentation. The situation would have to be carefully analyzed, but it is submitted that when the principal's incapacity is directly caused by SDAT, the research is to gain knowledge about SDAT, and the research must use advanced stage, incompetent SDAT individuals, the attorney's consent would not violate either the duty to comply with the principal's instructions, or the duty to act in the principal's interest.

A concept most widely known as a "Ulysses contract,"[48] but also referred to in various settings as "future-oriented consent,"[49] and a "voluntary commitment contract,"[50] has received increasing recent attention. Ulysses instructed his crew to bind him to the mast of his ship before they sailed past the irrestible Sirens and to ignore his requests—while he was under the irresistible influence of the Sirens' song—to be released or to sail nearer. A so-called Ulysses contract is a statement made by a competent individual who suspects that strange, mind-altering forces may come into play, and who explicitly instructs other parties to ignore his or her future decisions made under such unnatural influences. Thus a competent individual with SDAT may know that the progression of SDAT will cause growing incompetence. But that individual may desire to make some type of binding commitment, while competent, to continue to participate in mixed therapeutic and nontherapeutic SDAT procedures or in nontherapeutic SDAT procedures after he or she has become incompetent, and may also desire to consent in advance to begin these procedures after the disease has advanced beyond the point when he or she is still competent. With the help of the physician, the patient advocate, and others the patient could write a statement of his or her present, competent desire to participate when no longer competent to consent to do so. The statement could set forth the motives for giving such prior consent, the procedures that the patient wishes to be continued, the procedures that can be commenced when no longer competent, and circumstances under which his or her prior consent should be considered to be revoked.

The argument for the validity and enforceability of a Ulysses contract is that respect for the individual's person autonomy supports ignoring the incompetent refusal or withdrawal, and honoring the decisions that most accurately reflect his or her "true" wishes. Obviously we should proceed very slowly to sanction a binding consent that allows researchers to ignore a SDAT victim's attempts to refuse to participate or to withdraw from a project. If we are to consider compelling a nonassenting or objecting incompetent individual to participate in medical procedures upon the basis of prior consent, we must provide an elaborate system for protecting the individual's vulnerability.

A Tentative Model

I will here take the National Commission's recommendations as they affect research with incompetent SDAT individuals and offer a tentative model of proxy decision-making that I believe not only will satisfy court challenge, but will also serve as a firm foundation for fashioning a public policy in this increasingly important area. I will briefly identify my own views and my differences from the National Commission's views. My model is offered for those cases in which the research cannot be done without using incompetent SDAT individuals. Only if the particular research cannot be done without using incompetent SDAT individuals should they be asked to participate. Thus the proxy consent model is limited to a rather narrow area of research. In addition, this model is further limited to those cases in which the mental incapacity is a direct product of SDAT and SDAT is the disease being studied. I would modify the Commission's position in the cases involving more than minimal risk where the Commission allows the individual to assent and the court appointed guardian, if any, to give permission. In cases where there is no court appointed guardian, I would require that, in addition to the individual's assent, relevant family members give permission. This would take care of cases where no guardian is appointed because, although the individual assents, there is fear that a third party appointed as a guardian would not give permission for the individual to participate. Thus, under my proposal, if no guardian is appointed, relevant family members would have to give permission. Thus the model attempts to balance the value of protecting vulnerable individuals and the value of diagnosis, treatment, cure, and prevention of the very disease that causes the mental incapacity.

My biggest departure from the National Commission's approach occurs in those instances where the incompetent individual does not assent or object to either therapeutic procedures or mixed therapeutic and nontherapeutic procedures. There are specific instances in which the Commission would request a court to override the individual's objection. I do not provide for a procedure for the court to authorize an incompetent SDAT individual's participation in a research project, whether or not the individual objects to the procedure. In light of the instances in which I allow the individual himself, a family member, a guardian, or a durable power of attorney to either consent to participation or to authorize participation, I do not believe that court authorization for an incompetent's participation in a research project (1) provides a sufficient protection of the individual's person autonomy, or (2) is necessary to protect society's interest in attacking SDAT.

In only two instances do I think that it is permissible to compel a non-assenting or objecting incompetent individual to submit to mixed therapeutic and nontherapeutic SDAT procedures that involve more than minimal risk or to nontherapeutic procedures. It seems to me allowable to do so in the case in which, prior to becoming incompetent, and in accordance with state law, the individual has appointed a "durable power of attorney" that he or she wishes to begin or retain authority to act when the individual loses capacity to act for him or herself. It seems appropriate to allow, but not to require, the durable attorney to consent to procedures even though the incompetent individual does not or cannot assent to, or even though he now objects. The individual, while competent, has selected a person to act for him and make these types of decisions when he no longer has capacity to do so. One of the situations that the principal with SDAT can be assumed to have considered is that, once incompetent, he or she would fail to assent or would object to procedures that the competent self would have consented. The specific act of selecting a durable attorney to act in his or her stead in such instances can be taken as authorization for the attorney to act, even though the incompetent SDAT victim fails to assent or manifests an objection.

Although a great deal of analysis needs to be done, I think that it may also be permissible to compel a non-assenting or objecting incompetent individual to submit to mixed therapeutic and nontherapeutic SDAT procedures that involve more than minimal risk, or to nontherapeutic SDAT procedures that involve only a minor increase over minimal risk in cases where a Ulysses contract has been executed by the individual while competent. If prior to becoming incompetent, the individual has executed a statement specifying in advance the desire to make a binding commitment, under what conditions this agreement becomes binding, under what conditions the consent is to be revoked, which specific procedures are to be continued after incompetency is established, which specific procedures are to be begun upon becoming incompetent, and which of his or her then incompetent objections are to be respected or disregarded. The IRB's subject advocate should monitor both the individual's initial consent and subsequent behavior in order to determine whether the terms of the Ulysses contract are being fulfilled. In addition, a modified Ulysses contract might be used in which subsequent incompetent attempts to withdraw from the procedures could be overriden, but only after review by the subject advocate, an independent psychiatrist, or a hospital ethics committee. There are several troublesome points in the Ulysses contract procedure that must be carefully attended to. We must ascertain as best we can when the initial consent represents the individual's true wishes. We need to examine pressures or influences imposed upon the individual. Were his or her thought processes affected by psychotropic medication? Was the motive a fear of being abandoned by the medical care team after he or she becomes incompetent? The individual must remain free to competently refuse to make a Ulysses contract. But how do we distinguish competent refusal from incompetent refusal, and how do we further distinguish justifiable incompetent refusal (pain, discomfort, nausea) from unjustifiable incompetent refusal? However, there appears to be some room in appropriate situations to allow an individual to give a binding consent in advance when that individual knows that SDAT will cause him or her to become incompetent and that the procedures to which he or she specifically consents are of vital importance for understanding SDAT, may benefit future SDAT victims, or are conducted in the most dignified and humane manner possible.

Use of durable powers of attorney or Ulysses contracts to authorize an incompetent SDAT individual to participate despite his or her lack of assent, or the objection of the incompetent individual, should be limited to (1) therapeutic procedures and mixed therapeutic and nontherapeutic procedures with no more than minimal risk; (2) therapeutic procedures and mixed therapeutic and nontherapeutic procedures involving more than minimal risk, but which hold out the prospect of a direct benefit or a monitoring procedure required for well-being, when the risk is justified by anticipated benefit and the relation of risk to anticipated benefit is as favorable as any alternative; and (3) nontherapeutic procedures in which, although there is no direct benefit, the anticipated knowledge is of vital importance in understanding SDAT or reasonably expected to benefit future SDAT victims. In this last instance, I go beyond what the National Commission would allow. My model increases the role of institutional review boards generally, makes use of the newly emerging idea of a patient advocate working through the institutional review board, and utilizes the new concepts of durable powers of attorney and Ulysses contracts.

My model contains four major categories. For each category, after identifying the factual assumptions, I set out procedures that I believe will satisfy both legal and moral requirements.

1. The SDAT victim is *de facto* incompetent or *de jure* incompetent by court adjudication and the court has not appointed a guardian of the person. There has been no prior appointment of a durable power of attorney or execution of a Ulysses contract.
   a. The research project has been approved by the appropriate institutional review board for (1) the merits of the study, and (2) the appropriateness of using this particular SDAT individual.
   b. The institutional review board's patient advocate determines that the individual is incompetent to make decisions about participation in this research and that the proxy is an appropriate person to make a substituted judgment decision for the individual.
   c. The SDAT individual cannot be a research subject if he or she objects, regardless of the reason for the objection.
   d. Proxy permission, where permitted, can be given by the competent spouse, or if there is no competent spouse, then by the competent adult children, or if there are no competent adult children, then by the competent parents; or if there are no competent parents, then by the competent adult siblings.
   e. The researcher must disclose information to the proxy and obtain the proxy's permission under applicable rules of informed consent decision making.
   f. The appropriate proxy can give permission for the SDAT individual for mixed therapeutic and nontherapeutic procedures with no more than minimal risk and (1) an intervention that holds out the prospect of direct benefit to the SDAT individual, or (2) a monitoring procedure that is required for the individual's well-being, as determined by the institutional review board's patient advocate and the individual's independent physician.
   g. The appropriate proxy can give permission for the SDAT individual for mixed therapeutic and nontherapeutic procedures, holding out the prospect of direct benefit to the subject or a monitoring procedure required for the subject's well-being, involving more than minimal risk, where that risk is justified by the anticipated benefit to the subject as determined by both the institutional review board's patient advocate and the individual's independent physician.
   h. No proxy permission can be given for the SDAT individual for nontherapeutic procedures that involve only

a minor increase over minimal risk, but do not hold out either the prospect of direct benefit to the subject or a monitoring procedure that is not required for the subject's well-being, even though the anticipated knowledge either is (1) of vital importance for the understanding or amelioration of SDAT, or (2) may be reasonably expected to benefit future SDAT victims, as determined by either the institutional review board's patient advocate or the individual's independent physician.

2. The SDAT victim is a *de jure* incompetent by court adjudication and the court has appointed a guardian of the person or a guardian ad litem for this decision. There has been no prior appointment of a durable power of attorney or execution of a Ulysses contract.
   a. The research has been approved by the appropriate institutional review board for (1) the merits of the study, and (2) the appropriateness of using this particular SDAT individual.
   b. The SDAT individual cannot be a research subject if he or she objects, regardless of the reason for the objection.
   c. The researcher must disclose information to the guardian and obtain the guardian's permission under applicable rules of informed consent decision-making.
   d. The guardian can give permission for the SDAT ward for mixed therapeutic and nontherapeutic procedures with no more than minimal risk holding out the prospect of direct benefit to the SDAT individual, or a monitoring procedure that is required for the individual's well-being, as determined by the institutional review board's patient advocate and the individual's independent physician.
   e. The guardian can give permission for the SDAT ward for mixed therapeutic and nontherapeutic procedures, holding out the prospect of direct benefit to the subject or a monitoring procedure required for the subject's well-being, involving more than a minimal risk, where that risk is justified by the anticipated benefit to the subject, as determined by both the institutional review board's patient advocate and the individual's independent physician.
   f. The guardian cannot give permission for the SDAT ward for nontherapeutic procedures that involve only a minor increase over minimal risk, but do not hold out either the prospect of a direct benefit to the subject or a

monitoring procedure that is not required for the subject's well-being, even though the anticipated knowledge either is (1) of vital importance for the understanding or amelioration of SDAT, or (2) may be reasonably expected to benefit future SDAT victims, as determined by either the institutional review board's patient advocate or the individual's independent physician.

3. The SDAT victim is *de facto* incompetent or *de jure* incompetent by court adjudication. There has been a prior appointment of a durable power of attorney in accordance with applicable state law, but no execution of a Ulysses contract.
   a. The research project has been approved by the appropriate institutional review board for (1) the merits of the study, and (2) the appropriateness of using this particular SDAT individual.
   b. The institutional review board's patient advocate determines that the SDAT individual is incompetent according to the express terms of the durable power of attorney.
   c. The researcher must disclose information to the durable attorney and obtain the attorney's permission under applicable rules of informed consent decision-making.
   d. The durable attorney can consent for the SDAT individual for mixed therapeutic and nontherapeutic procedures with no more than minimal risk holding out the prospect of direct benefit to the SDAT individual, or a monitoring procedure that is required for the individual's well-being, as determined by the institutional review board's patient advocate and the individual's independent physician.
   e. The durable attorney can consent for the SDAT individual for mixed therapeutic and nontherapeutic procedures holding out the prospect of direct benefit to the subject or a monitoring procedure required for the subject's well-being, involving more than a minimal risk, when that risk is justified by the anticipated benefit to the subject as determined by both the institutional review board's patient advocate and the individual's independent physician.
   f. The durable attorney can consent for the SDAT individual for nontherapeutic procedures that involve only a minor increase over minimal risk, but do not hold out

either the prospect of a direct benefit to the subject or a monitoring procedure that is not required for the subject's well-being, but the anticipated knowledge either is (1) of vital importance for the understanding or amelioration of SDAT, or (2) may be reasonably expected to benefit future SDAT victims, as determined by either the institutional review board's patient advocate or the individual's independent physician.

4. The SDAT victim is *de facto* incompetent or *de jure* incompetent by court adjudication. There has been a prior execution of a Ulysses contract, but no execution of a durable power of attorney.
   a. The research project has been approved by the appropriate institutional review board for (1) the merits of the study, and (2) the appropriateness of using this particular SDAT individual.
   b. The institutional review board's patient advocate determines that the SDAT individual is incompetent according to the express terms of the Ulysses contract.
   c. The Ulysses contract can authorize for the SDAT individual mixed therapeutic and nontherapeutic procedures with no more than minimal risk holding out the prospect of direct benefit to the SDAT individual, or a monitoring procedure that is required for the individual's well-being, as determined by the institutional review board's patient advocate and the individual's independent physician.
   d. The Ulysses contract can authorize for the SDAT individual mixed therapeutic and nontherapeutic procedures holding out the prospect of direct benefit to the subject or a monitoring procedure required for the subject's well-being, involving more than a minimal risk, where that risk is justified by the anticipated benefit to the subject as determined by both the institutional review board's patient advocate and the individual's independent physician.
   e. The Ulysses contract can authorize for the SDAT individual nontherapeutic procedures that involve only a minor increase over minimal risk, but do not hold out either the prospect of a direct benefit to the subject or a monitoring procedure that is not required for the subject's well-being, but the anticipated knowledge either is (1) of vital importance for the understanding or amelioration of SDAT, or (2) may be reasonably expected to benefit future SDAT victims, as determined

by either the institutional review board's patient advocate or the individual's independent physician.

There are, of course, many important and difficult issues that cannot be analyzed here. The extremely difficult issue of determining when an individual is *de facto* incompetent to decide particular important life decisions in particular circumstances in those cases where there has been no judicial determination of incompetence cannot be adequately addressed here. Particularly difficult are those cases, common in SDAT, where the individual's mental condition varies from day to day and can change dramatically in a short period of time. I will not attempt to define more specifically therapeutic procedures, mixed therapeutic and nontherapeutic procedures, and nontherapeutic procedures. If the individual has executed a durable power of attorney and a Ulysses contract, I would give priority to the durable attorney in the event that he or she refused to allow the incompetent SDAT individual to participate. But I will not here set out the rationale for doing so. Finally, the paper does not discuss the power of courts to authorize or compel an individual to submit to mixed therapeutic and nontherapeutic procedures.

Conclusion

The issues involved in the use of derived consent in SDAT research represent a conflict between society's obligation to prevent suffering by developing methods of prevention, diagnosis, treatment, and cure of SDAT, and society's obligation to avoid exploiting the vulnerability—and to promote the personal autonomy—of SDAT victims. Much additional work awaits us in clearly identifying the specific types of research activities in which we can morally ask those suffering from SDAT to participate. We must continually identify and analyze conflicting value choices. The dilemmas we face in questions of proxy consent to participate in this important area of research force us to squarely confront the uncomfortable task of assigning priorities to several deeply held human values when they conflict. This task, agonizing though it may be, should instill within each of us individually, and within society, generally, a renewed respect for the dignity of every individual.

Notes and References

[1]"Where the complaint in suit is unauthorized treatment of a patient legally or factually incapable of giving consent, the established rule is that, absent an emergency, the physician must obtain the necessary authority from a relative." *Canterbury v. Spence*, 464 F.2d 772, 789 n.92 (DC Cir.), *cert. denied* 409 US 1064 (1972).

"Where a patient has been legally adjudged incompetent, consent to treatment must be secured from the patient's legal guardian. Where no judicial determination of incompetency has been made, the consent of the spouse or parent should be obtained where incompetency is reasonably suspected." King, the Law of Medical Malpractice (1977) at 140.

"The accepted practice is to get the consent of the patient himself if he is a competent adult, or if this is not possible, to obtain the consent of a spouse, parent, relative, or someone in a capacity as a guardian. If, of course, another person gave consent for the patient plaintiff while the patient plaintiff was unable to give consent, then this would be a proper defense to an action brought by a plaintiff patient alleging that either no consent or improper consent was given by him for the medical treatment or surgical operation." Annot., 25 A.L.R.3d 1441 (1969).

[2]See, e.g., *Schloendorff v. Society of New York Hospital*, 211 NY 125, 105 N.E. 92, (1914); *Salgo v. Leland Stanford Jr. University Board of Trustees*, 154 Cal. App.2d 560, 317 P.2d 170 (1957); *Natanson v. Kline*, 186 Kan. 393, 350 P. 2d 1093, *opinion on denial of motion for rehearing*, 187 Kan. 186, 354 P.2d 670 (1960); *Canterbury v. Spence*, 464 F.2d 772 (DC Cir.), *cert. denied* 409 US 1064 (1972); *Cobbs v. Grant*, 8 Cal.3d 229, 104 Cal. Rptr. 505, 502 P.2d 1 (1972); President's Commission for the Study of Ethical Problems in Medicine and Biomedical and Behavioral Research, *Making Health Care Decisions* (1982).

[3]Annas, Glantz, & Katz, Informed Consent to Human Experimentation (1977) at 153 to 154 and 156.

[4]Katz and Capron, *Catastrophic Diseases: Who Decides What?* (1975) at 108 to 109.

[5]*Ibid.* at 154 to 155.

[6]Fried, *Medical Experimentation: Personal Integrity and Social Policy* (1974) at 25 to 26.

[7]*Ibid.* at 28 to 29.

[8]*Ibid.* at 29 to 30.

[9]*Ibid.* at 26 to 27.

[10]Ark. Stat Section 82–363(b) (Cum. Supp. 1979).

[11]Ohio Rev. Code Ann. Section 2317.54(c) (Page 1981).

[12]Idaho Code Section 39–4302 (1980).

[13]Mississippi Code Ann. 1972 Section 41–41–3(e); Ark. Stat. Section 82–363(3) (Cum. Supp. 1979).

[14]North Carolina Gen. Stat. Section 90–21.13 (1981).

[15]Ohio Rev. Code Ann. Section 2317.54(C) (Page 1981).

[16]Idaho Code Section 39–4303 (1980).

[17]North Carolina Gen. Stat. Section 14–22.2 (Cum. Supp. 1981).

[18]Tenn. Code Ann. Section 33–307 (Cum. Supp. 1977).

[19]Alaska Stat. Section 47.30.130(b) (1979).

[20]Conn. Gen. Stat. Ann. Section 17–206d(c) (Cum. Supp. 1981).

[21]NJ Stat. Ann. 30:4–72(c) (West 1981).

[22]*See,* Calif. Welfare & Institutions Code Section 5326.6 (psychosurgery) and Section 5326.7 (electroconvulsive therapy) (West Supp. 1980).

[23]Calif. Welfare & Institutions Code Section 5326.6 and Section 5326.7 (West Supp. 1980).

[24]See, e.g., *In re Quinlan,* 70 NJ 10, 355 A.2d 647, *cert. denied* sub. nom., *Garger v. New Jersey,* 429 US 922 (1976); *Superintendent of Belchertown State School v. Saikewicz,* 373 Mass. 728, 370 N.E.2d 417 (1977); *Matter of Spring,* Mass. Adv. Sh. 1209, 405 N.E.2d 115 (1980); *Eichner v. Dillon,* 438 N.Y.S.2d 266, 420 N.E.2d 64 (1981).

[25]*See,* for example, Calif. Health & Safety Code Section 7185–7195 (West Ann. Supp. 1981); Wash. Rev. Code Ann. Section 70.122.010–70.122.905 (1981); President's Commission for the Study of Ethical Problems in Medicine and Biomedical and Behavioral Research, *Deciding to Forego Life-Sustaining Treatment* (1983).

[26]Annas, *supra* note 3 at 155 to 156.

[27]*McFall v. Shimp,* 10 Pa. D. & C.3d 90 (Pa. 1978). In a more recent case, a woman's blood had been tissue-typed by the University of Iowa Hospital to determine if she would be a suitable bone marrow donor for her child. The woman, referred to as "Mrs. X" was routinely listed in the hospital's bone marrow transplant registry. A Leukemia Victim, whose prognosis without a bone marrow transplant was grim, asked the court for a mandatory injunction to require the hospital to disclose Mrs. X's name either to the court or to his attorney. He proposed that the court or counsel then be permitted to write Mrs. X to notify her of his need and her possible suitability as a donor, asking her if she would consider donating bone marrow to him to save his life. The trial court ordered the hospital to send a letter to Mrs. X. The Iowa Supreme Court reversed on the narrow grounds that although the bone marrow donor registry is a public record under the state's public record statute, a person who submits to tissue typing tests to determine suitability as a donor is a "patient," and records of the tissue typing are confidential hospital records, exempt by the statute from the general public's disclosure rights. The decision shields those who have been tissue typed from being beseiged by pleas from leukemia and aplastic anemia victims to undergo further tests to determine suitability to donate lifesaving bone marrow to them. *Head v. Colloton,* 331 N.W.2d 870 (Iowa 1983).

[28]Fried, *supra* note 6 at 23.

[29]Annas, *supra* note 3 at 172.

[30]Katz and Capron, *supra* note 4 at 109.

[31]Annas, *supra* note 3 at 174; Katz and Capron, *supra* note 4 at 109.

[32]Annas, *supra* note 3 at 175.

[33]*Ibid.* at 179.

[34]*Ibid.* at 182 to 183.

[35]Capron, Informed Consent in Catastrophic Disease Research and Treatment, 123 U. Pa. L. Rev. 340 at 417 to 428 (1974).

[36]Katz and Capron, *supra* note 4 at 112.

[37]*See,* National Commission for the Protection of Human Subjects of Biomedical and Behavioral Research, Report and Recommendations on Research Involving Children, DHEW Publication No. (OS) 77–0004 (1977); National Commission for the Protection of Human Subjects of Biomedical and Behavioral Research, Report and Recommendations on Research Involving Those Institutionalized as Mentally Infirm, DHEW Publication No. (OS) 78–006 (1978).

[38]Report and Recommendations on Research Involving Those Institutionalized as Mentally Infirm, *supra* note 37, at 7 to 8.

[39]*Ibid.* at 11 to 12.

[40]*Ibid.* at 16 to 17.

[41]Robertson, "Ten Ways to Improve IRBs," *Hastings Center Report* 9, February 1979, 29.

[42]McGrath & Briscoe, "The Role of the Subject Advocate in a Community-Based Medical Research Facility, *IRB: A Review of Human Subjects Resarch* 3, March 1981, 7.

[43]Robertson, *supra* note 41 at 31.

[44]President's Commission for the Study of Ethical Problems in Medicine and Biomedical and Behavioral Research, Deciding to Forego Life Sustaining Treatment (1983), supra note 25.

[45]Ballentine's Law Dictionary 971 (3d Edition 1969).

[46]3 Am. Jur. 2d *Agency* Section 23–33 (1981).

[47]Uniform Probate Code (U.L.A.) sections 5-501 to 5-503 (revised 1979) and Uniform Durable Power of Attorney Act (U.L.A.). The nine states that have adopted the Uniform Durable Power of Attorney Act are: Ala. Code 1975, section 26-2-2 (1983); West's Ann. Cal. Civ. Code, sections 2400 to 2407 (1983 Cum. Supp.); 12 Del. Code sections 4901 to 4905 (1982 Cum. Supp.); Ida. Code sections 15-5-501 to 15-5-507 (1983 Cum. Supp.); Kan. Stat. Ann. sections 58-610 to 58-617 (1983); Mass. Gen. Laws Ann. Ch. 201B, sections 1 to 7 (1983); Vernon's Ann. Mo. Stat. sections 486.550 to 486.595 (1984); Tenn. Code Ann. sections 34-13-101 to 34-13-108 (1983 Supp.); Wis. Stat. Ann. section 243.07 (1983). Some state statutes providing that a power of attorney may continue after a disability predate the Uniform Acts, *e.g.*, Pa. Stat. Ann. tit. 20 section 5601 (Purdon 1975) and N.Y. Gen. Oblig. Law section 5-1601 (Consol 1978).

In addition, California has enacted a separate set of provisions relating to durable power of attorney for health care. West's Ann. Cal.

Civ. Code, sections 2430 to 2443 (1983 Cum. Supp.). Several other state legislatures are considering such a statute.

[48]"Elster, Ulysses, and the Sirens: Studies in Rationality and Irrationality" (1979); "Case Studies: Can a Subject Consent to a 'Ulysses Contract'?," *Hastings Center Report* 12, August 1982, 26.

[49]Wexler, Therapeutic Justice, 57 Minn L. Rev. 289 (1972), reprinted in National Commission on Marijuana and Drug Abuse, Drug Use in America: Problem in Perspective, app. IV, at 560 (1973).

[50]Howell, Diamond, & Wikler, "Is There a Case for Voluntary Commitment?" in Beauchamp & Walters, Eds., *Contemporary Issues in Bioethics* (2d Ed. 1982); Dresser, Ulysses and the Psychiatrists: A Legal and Policy Analysis of the Voluntary Commitment Contract, 16 Harv. C.R.-C.L. L. Rev. 777 (1982).

Clinical Research in Senile Dementia of the Alzheimer's Type

Suggested Guidelines Addressing the Ethical and Legal Issues

Vijaya L. Melnick, Nancy Dubler, Alan Weisbard, and Robert N. Butler

In November 1981, the National Institute on Aging (NIA) sponsored a conference on "Senile Dementia of the Alzheimer's Type (SDAT) and Related Diseases: Ethical and Legal Issues Related to Informed Consent." The conference was convened to explore the values, conflicts, and competing interests that must be accommodated if research on the pathophysiologic processes and psychosocial aspects of dementia is to continue.

The papers presented at the conference and the discussions that followed pointed to a clear need for more research. However, what type of research might be acceptable, in the context of present Federal regulations governing research on human subjects, was not at all clear. Questions were raised as to:

1. Who might or should be permitted to speak on behalf of or as a substitute for the patient when the patient is not capable of giving competent informed consent, SDAT patients most often being in a state of progressively declining competence.
2. How, in the most ethical context, could protocols be designed that are scientifically acceptable and that will permit research to further the elucidation and treatment of dementing illnesses.
3. Whether patients afflicted with such diseases can ever grant effective informed consent to participate in a research protocol.

4. Whether patients can ever be morally admitted to protocols in the absence of their own consent or the adequate consent of others, and how the roles of the family and close friends are defined in this context.
5. What the possibilities are for obtaining consent at an early stage of the illness for later research intervention.
6. What additional constraints are placed on researchers attempting to address the complex problems of dementing illnesses.
7. Whether society can make a claim, however minimal, on the supposed altruism of patients if the intervention poses minimal risk and the possible benefit is great.

In an attempt to answer these questions to the extent possible, the NIA sponsored a task force to design a set of guidelines that might be of help to researchers, as well as to those who formulate and review research protocols concerning SDAT or involving SDAT patient subjects. In addition, it should be of some benefit to the patients and their families who might be involved with such research.

What follows is a set of suggested guidelines that resulted from the deliberations of the task force committee.

Senile dementia of the Alzheimer type (SDAT) devastates millions of afflicted persons and their families. The disease limits and progressively destroys the capacity of individuals to function competently and to live independently; results in massive disruptions of family life; and, in its advanced stages, often necessitates permanent institutionalization. At present, SDAT is difficult to diagnose, impossible to cure or arrest in its course, and poorly understood. These human costs of SDAT provide a compelling justification for research directed to achieve a better understanding of the disease and eventually bring it under control.

Several aspects of SDAT, although not unique to that disease, give SDAT research (defined here as research on SDAT-afflicted subjects regarding their disease) a special character among classes of research involving human subjects. First, because of the interconnections of SDAT with human consciousness, memory, and cognitive and affective abilities, animal research is particularly limited as a source of improved understanding of the disease or as a step in progress toward effective treatment. Thus, if research is to proceed, human subjects must be employed even during the early stages of seeking a basic scientific understanding of SDAT.

Second, patients afflicted with SDAT are, virtually by definition, on a path of declining competence. Early in the course of their disease, when memory difficulties and impaired word-finding skills are the predominant deficits associated with the disease, many SDAT patient–subjects will retain the capacity to provide valid consent (that is, consent meeting ethical and legal requirements) to participation in research. As the disease progresses, however, this capacity is progressively impaired and ultimately extinguished. Thus, patients with advanced disease lack the capacity to consent on their own behalf to participation in research. If certain types of research requiring the participation of patients with advanced disease are to go forward, the participation of such patients as research subjects must be justified on some basis other than their personal and contemporaneous informed consent.

Third, because of the debilitating nature of SDAT, many patients with advanced disease must be institutionalized in nursing homes or other chronic care facilities. Although some such institutions have exemplary records of care and concern, others have been shown to have acted in ways contrary to the best interests of their patients. In light of the abuses that have taken place within the all-too-recent past, the conduct of research on incompetent (defined here as those without capacities to give or withhold consent) and institutionalized patients properly evokes special sensitivities and requires especially careful scrutiny.

These three special characteristics suggest that a model for regulating research predicted on an assumption that potential subjects are competent adults possessing the capacity to evaluate for themselves the potential risks and benefits of participation in research, as well as the capacity to withdraw from participation during the course of the research if they so choose, may be ill-suited to the different constellation of facts present in SDAT research. Existing Federal regulations provide only limited guidance, and efforts to address such issues more specifically, such as the very useful reports of the National Commission for the Protection of Research Subjects of Biomedical and Behavioral Research concerning research on children and on those institutionalized as mentally infirm, have not yet resulted in needed modifications to the existing Federal regulations.

In the absence of such modifications, the application of existing Federal regulations to SDAT research poses a number of troubling ethical and legal issues. From an ethical and policy perspective, society must resolve whether and under what conditions to permit research, absent the personal and contemporaneous informed consent of the research subject, which may not be directly beneficial to the subject but that is expected to increase knowledge and ultimately to assist in diagnosis, prevention, amelioration, or cure. The interest of present and future generations in bringing SDAT under control must be balanced against the daunting recognition that efforts to achieve this much desired end may impinge on the individual rights or welfare of a particularly vulnerable group, i.e., SDAT patient subjects who lack the basic abilities to understand, appreciate, and consent to participation in research. From the legal perspective, both the existing Federal regulations and, perhaps more significantly, the laws of many states raise substantial doubts as to the legal capacity of SDAT patient–subjects, or of others who may act on their behalf, to provide legally effective consent to participation in certain types of research. Fundamentally, society must consider whether the doctrine of informed consent provides an adequate mediating principle for the resolution of these troubling issues and indeed whether recourse to requirements for personal informed consent makes sense when the individual most concerned—the potential subject—although "willing" to participate in research, lacks the capacity to provide a "competent" informed consent.

The guidelines that follow are intended to bring to the attention of those who formulate or review research protocols involving SDAT patient–subjects some of the specific issues and concerns that attend such research in the present legal and ethical climate. The guidelines are educational rather than regulatory in purpose; they are intended to stimulate discussion rather than terminate it. Investigators and institutional review boards (IRBs) are specifically cautioned that compliance with these guidelines is not a substitute for compliance with applicable Federal regulations and state laws.

The guidelines are organized to provide guidance and assistance to IRBs and to researchers with respect to the following issues:

1. to express a preference for research with patients who are competent or who are otherwise relatively less vulnerable to potential abuse;
2. to identify individuals who are favorably inclined to participation in research and to provide mechanisms for their

participation now and in the future, subject to necessary safeguards;

3. to assure that all research protocols involving SDAT patient–subjects have adequate mechanisms to assess competence, assure the adequacy of the consent process, and assure the continued ability of the subjects to decline to participate or to withdraw;
4. to indicate special considerations in and limitations on research involving patients who are not capable of granting legally effective consent on their own behalf.

Suggested Guidelines

1. *Selection of subjects for inclusion in SDAT research*

   When not incompatible with the fundamental objectives of the research, research involving SDAT patients should be designed to enroll as subjects in preferred descending order:

   a. Non-institutionalized SDAT patient–subjects who retain the capacity to decide for themselves, on a competent basis, whether or not to participate in research.
   b. Non-institutionalized SDAT patient–subjects of impaired capacity who previously (while competent) expressed a willingness to participate in research, continue to express such willingness, and who retain an active and caring family supportive of the individual's participation in research.
   c. Non-institutionalized SDAT patient–subjects of impaired capacity who express a current willingness to participate in research and who retain an active and caring family supportive of the individual's participation in research.
   d. Other non-institutionalized SDAT patient–subjects of impaired competence who retain some ability to care for themselves and who express a current willingness to participate in research.

2. *Qualification for institutional sites*

   In the event that research on institutionalized patients is justified, procedures should be established by an IRB to ensure that the clinical and supportive care provided by the institution is of acceptable quality. "Acceptable quality" requires, *inter alia,* that institutions comply fully with all Federal and state regulations and requirements that govern levels of care, staffing, and the protection of patients' rights. If particular institutions do not meet the standards, the IRB should refuse permission for research to be initiated or continued in those settings. Among institutionalized patients, priorities for selection of research subjects should be in general accord with the hierarchy indicated in guideline 1.

3. *Encouragement for "durable powers of attorney"*

   Researchers should be encouraged to develop and to employ documents such as "durable powers of attorney," or other empowerments permitted by specific state statute to survive disability, by which a competent patient may appoint an agent (often, but not necessarily, a family member) with specific power to consent to and supervise the continued participation of the patient in a research protocol. Such documents may also include substantive guidance to the agent, indicating the individual's attitudes and desires regarding participation in research. The IRBs should be attentive to the status of such documents under state law, and should discourage documents naming members of the research team as agents for these purposes.

4. *Encouragement for long-range/long-term protocols*

   Researchers on SDAT should be encouraged to devise long-range, long-term protocols for which valid consent could be obtained from patient–subjects with early diagnoses, to permit the subsequent participation of these patient–subjects when their capacity to provide contemporaneous valid consent is diminished or extinguished. This guideline is particularly appropriate for protocols that propose continuing observation and frequent reevaluation of patient–subjects.

5. *Determination of particular subject's capacity for a specific protocol*

   In the event that a particular research project can proceed only with subjects whose capacity to provide or to withhold legally effective informed consent is likely to be open to substantial question, the IRB should assure that the research protocol provide adequate means for determining the capacity of potential subjects. Each research protocol should specify the means of evaluation to be employed. Determinations of capacity must be made on an individualized basis and should focus on the ability of the potential subject:

   a. To understand the nature and consequences of the particular participation in research.

b. To comprehend the fact that the suggested intervention is in fact research (and is not intended to provide therapeutic benefit to the subject, when that is the case).
c. To comprehend that alternatives exist, including the alternative not to participate, without jeopardizing the care and concern of health-care providers.

The determination of the subject's capacity to consent to participation in research should not be dependent upon an assessment of the subject's "overall state of competency." Thus, an individual may retain the capacity to consent to participation in research, but lack the capacity, for example, to manage his or her financial affairs, or vice versa. However, IRBs should recognize that persons legally adjudicated as incompetent may not provide legally effective consent to participation in research.

6. *Relationship of capacity to consent to potential risks of research*
The greater the risks posed by the research intervention, the lesser the direct benefits likely to accrue to the subject from participation in the research, and the more complex the procedures involved in the research, the greater the need for careful scrutiny of the potential subject's capacity to provide consent. In order to assure such careful scrutiny, IRBs should consider, and may require, that specific mechanisms be employed to review the determinations of capacity and the adequacy of consent provided by subjects who may lack the capacity to consent. Such mechanisms may include the use of "two-part consent processes" (the first part to assure the subject's capacity to understand, and the second part to monitor the actual understanding of the issues involved in the proposed research), and independent evaluators of subject capacity.

7. *Subject's right to object to participation*
Except for one limited category of potential exception, no research intervention should be commenced upon any subject (whether or not competent) who objects to participation in the research, nor continued with any subject who objects to participation in the research, without regard to whether the patient–subject earlier provided consent to participation. The only potential exception to this rigorous prohibition arises when an intervention of significant therapeutic potential to the patient is legitimately available solely in connection with participation in research, when the provision of this intervention is clearly in the subject's best interests and would be ordered in a

purely therapeutic context, and when any additional safeguards deemed appropriate are satisfied.

The IRB should consider, and may require, that provision be made for a "research auditor" who is able to maintain surveillance over the process and progress of research and to assure that the changing condition and perceptions of the patient with regard to continued participation in the research will receive immediate and sufficient regard and respect. This person shall have the right to withdraw a subject from participation in any protocol. In light of this guideline, researchers should be encouraged to design protocols in which substantial numbers of withdrawals will not prejudice the results.

8. *Authorization by a "legally authorized representative"*

The IRB should ensure that when a determination is made that a potential patient–subject lacks the capacity to provide valid consent to participation in research, but does not object to so participating, the protocol presents clear procedures for the designation of a substitute decision-maker who shall be qualified to function as the patient's "legally authorized representative" under applicable state law and Federal regulations. Such substitute decision-makers should possess the following characteristics:

a. No evident or substantial conflict of interest that would be likely to lead to a decision contrary to the best interests of the patient.
b. An ability to participate in a vigorous, informed and conscientious manner in the decision.
c. An ability to remain a vigorous advocate of the incompetent's interest in maintaining control of decision-making throughout the course of the patient's participation in research.

9. *Enrollment of competent patients as research subjects*

For SDAT patients with the capacity to give or to withhold legally effective consent, the patient's informed consent shall be required for participation in research protocols, together with such additional safeguards as the IRB may find appropriate.[a]

[a]In SDAT research, as in all other research, certain categories of research, i.e., those in which the essential values protected by informed consent are not at issue, and in which there is no serious possible breach of confidentiality, may be conducted without the informed consent of the subject, although with the prior review and approval of an IRB. Such research would include but not be limited to retrospective record reviews or the methodical notations of regular caregivers in an organized fashion.

10. *Authorization of enrollment of subjects in research*

For those subjects who lack the capacity to provide valid contemporaneous informed consent to participation in research, and who have not previously executed a durable power of attorney or otherwise manifested valid consent to a protocol when capable of so doing, participation as research subjects raises serious questions of ethics and law. The IRB should be especially attentive in such cases to the adequacy of protocol design and methodology (since a poorly conceptualized study whose benefit is unlikely to be significant will justify scant risk) and particularly sensitive to issues of patient selection.

For purposes of ethical and legal analysis, classes of research involving SDAT patient–subjects without capacity to provide informed consent may be roughly grouped in three categories. The categories and the considerations appropriate to each that could permit the enrollment of such patient–subjects without the subject's consent are discussed below. These categories do not apply if the patient–subject objects to participation.[b] The categories are:

a. Non-intrusive, non-invasive data collection and observation, and invasive research interventions not posing more than "minimal risk" to subjects.[c] Such research may be approved provided that:

(1) the research protocol is otherwise acceptable to the IRB;
(2) the research is relevant to the subject's condition;
(3) the subject's legally authorized representative provides legally effective informed consent for the sub-

[b] The IRBs should be especially attentive to the provision of applicable state laws, which in many cases limit the capacity of family members and legally authorized representatives to provide legally effective consent on behalf of incompetent subjects to interventions that are not beneficial and that pose potential risks to the incompetent's well-being.

[c] "Minimal risk" is defined, as in the Federal regulations, to mean "that the risks of harm anticipated in the proposed research are not greater, considering probability and magnitude, than those ordinarily encountered in daily life or during the performance of routine physical or psychological examinations or tests." In applying this definition to SDAT patient–subjects, it should be remembered that many elderly individuals are especially averse to risks or discomforts that may be taken for granted by other members of the population, and these special sensitivities, when they exist, should be respected.

ject's participation in research consistent with the requirements of 45 CFR 46.116 (a) and (b), except where such consent is not required pursuant to 46.116 (c) or (d).[d]

b. Invasive research interventions posing more than minimal risk that offer some realistic possibility of direct therapeutic benefit to the subject. Such research may be approved provided that:
(1) the research protocol is otherwise acceptable to the IRB;
(2) the research is relevant to the subject's condition and holds the prospect of direct benefit for individual subjects;
(3) the risk is justified by the anticipated benefit to the subject;
(4) the subject's legally authorized representative provides legally effective informed consent for the subject's participation in the research, consistent with the requirements of 45 CRF 46.116. Such consent is analogous to that provided by family members for therapeutic interventions outside the research context, and is recognized under the law in many states to the extent the anticipated benefits justify the risks involved. As the risk–benefit ratio worsens, the legal status of such consent is increasingly problematic, and the considerations for IRB review tend to merge with those applicable to research in category c.

c. Invasive research interventions posing more than minimal risk that are not associated with the realistic possibility of direct therapeutic benefit to the subject. Research interventions in this category pose more than minimal risk to subjects witout any compensating expectation of direct therapeutic benefit to individual subjects. When conducted on subjects lacking the capacity to grant legally effective informed consent (and who have not executed a durable power of attorney or other clear and directive instrument), such research poses in the most dramatic form the conflict between the societal interest in the conduct of important and promising research and the interests of the potential subject; these interests are protected both by

[d]See 46.116 for General Requirements for Informed Consent.

the doctrine of informed consent and by the laws of many states that limit the ability of family members or legal guardians to authorize such interventions. From a legal perspective, authorization of such participation by incompetent individuals in this category of research would probably require, in many states, specific approval by a court of competent jurisdiction. It may be that in at least some instances, the importance of research may ethically justify interventions posing greater than minimal risk on a willing subject who lacks the capacity to grant legally effective informed consent, even in the absence of any realistic probability of direct therapeutic benefit to the subject. Such a conclusion must rest on a thoroughgoing assessment of the risks involved and of the scientific importance of the research. When the local IRB believes these conditions are met, the sensitivity and the public importance of the issue strongly indicate the advisability of further definitive review of the particular protocol by a national ethics advisory body whose decisions shall be made in the course of a public process. Investigators and IRBs should be aware that even such a national review may not resolve the legality of the consent under the law of a particular state and that advice should be sought from competent counsel in all such cases.

Epilog

These guidelines are meant to raise issues; not to dispose of them. Members of IRBs should note that guidelines are not law; these guidelines have *no power* to authorize extensions of state law in areas of authority to consent for the participation of a patient–subject in research.

Institutional review boards might want to encourage:

1. The development of Federal policy for minimal-risk research that might provide criteria to guide efforts at the state level in drafting legislation to determine who could serve as a legally authorized representative for consent in those circumstances.
2. The establishment of a national research ethics advisory body with authority to endorse or prohibit specific re-

search protocols. Endorsement would be evidence of compliance with Federal regulations.

Institutional review boards might want to incorporate some or all of the following suggestions in regard to consent forms and procedures:

1. In any situation involving SDAT patient–subjects, a properly constructed informed consent form can be very helpful to both researcher and subject.
2. Suggestions for improving standard informed consent forms currently in use are the following:

   a. They should be made more readable, i.e., easier to read and to comprehend. Some ways to do this are to keep the forms short; make the print large type, e.g., Orator 10 type; use words familiar to the subjects reading them (vernacular instead of medical jargon); have the forms critiqued by fellow subjects, especially SDAT patient–subjects, rather than fellow researchers.
   b. Researchers studying SDAT might employ a consent process that assesses the subject's actual understanding of the explanation of the protocol and thus identifies subjects who need more explanation (use of a two-part consent form).
   c. An advantage of the two-part consent form is the built-in delay necessitated by the later administration of the second part (which measures actual understanding). Such a delay means more time for SDAT patient–subjects to confer with family members, think of important questions, and consider decisions.
3. Some suggestions for improving the informed consent procedures for SDAT patient–subjects are as follows:

   a. Leave a copy of the informed consent form with the subject for leisurely reading and re-reading, and discussion with family or friends or patient surrogate.
   b. Encourage "significant others" to be present at the time of the explanation(s) of the protocol if the subject also wishes them to be present.
   c. Consider perceptual adjuncts such as a tape recording, videotape recording, or slides that the subject may refer to before and after consenting to the experiment. Such adjuncts may refer to the material discussed in the protocol or the interview itself.

Members of the Task Force

Vijaya L. Melnick, PhD
Department of Biology
University of the District of Columbia
4200 Connecticut Avenue, NW Bldg. 3
Washington, DC 20007
Taskforce Group Coordinator and Chair
Formerly:
Special Assistant for Policy and Bioethics/OPEA
National Institute of Aging

Robert N. Butler, MD
Department of Geriatrics and Adult Development
Mt. Sinai Medical Center
Annenberg Building
24-64 One Gustave L. Levy Place
New York, NY 10029
Formerly:
Director, National Institute on Aging

Nancy Dubler, LLB
Department of Social Medicine
Montefiore Medical Center
111 East 210 Street
Bronx, NY 10467

Robert Katzman, MD
Department of Neurology
Albert Einstein College of Medicine
1300 Morris Park Avenue
Bronx, NY 10461

Joann H. Lynn, MD
Division of Geriatric Medicine
George Washington School of Medicine
2000 K Street, NW
Washington, DC 20006

Charles R. McCarthy, PhD
Office of Protection from Research Risk
National Institutes of Health
Westwood Building, Room 3A-18
Bethesda, MD 20205

Bruce Miller, PhD
Medical Humanities Program
Michigan State University
A106 East Fee Hall
Michigan State University
East Lansing, MI 48824

Richard Ratzan, MD
87 Hunter Drive
West Hartford, CT 06107

Lewis T. Rowland, MD
Neurological Institute
Columbia School of Physicians and Surgeons
710 W. 168th Street
New York, NY 10032

Barbara Stanley, PhD
Department of Psychiatry
Lafayette Clinic
951 East Lafayette
Detroit, MI 48207

Alan Weisbard, JD
Cardoza Law School
Yeshiva University
New York, NY 10003
Formerly:
President's Commission for the Study of Ethical Problems
in Medicine and Biomedical Behavioral Research

Index